THE BOOK OF
NUMBERS

THE BOOK OF
NUMBERS

FROM ZERO TO INFINITY, AN ENTERTAINING
LIST OF EVERY NUMBER THAT COUNTS

TIM GLYNNE-JONES

SIRIUS

SIRIUS

This edition published in 2024 by Sirius Publishing, a division of
Arcturus Publishing Limited,
26/27 Bickels Yard, 151–153 Bermondsey Street,
London SE1 3HA

Art direction: Beatriz Reis Custodio
Design: Alex Ingr

ISBN: 978-1-3988-3974-8
AD000084US

Printed in China

CONTENTS

Introduction

The first we know of numbers is when we start learning to count. One, two, buckle my shoe... Pretty soon we know the number of our age, the number of the day we were born, the month, the year. Before long we've learnt the numbers we like on the remote control, our friends' telephone numbers, the number of our favorite football player, how much pocket money we're owed and the cost of the things we want to buy... In the space of a handful of years, our knowledge of numbers soars from one and two to thousands and millions. And it goes on growing ad infinitum.

Numbers have a magical quality. Some people claim to see certain numbers appearing everywhere they look and attach supernatural power to it. In mathematics too, the way some numbers behave can seem amazing. Even Pythagoras, the great Greek mathematician, attributed mystical qualities to some of the numbers that captured his imagination. In some cases, numbers have assumed cult status from their appearance in popular culture, religion, mythology or historical events: 9/11, Catch-22, Room 101, 666—the number of the beast.

Amidst all of this it's easy to forget that most of the numbers we use, and the ways they are applied, are the invention of man. That there are 24 hours in a day, and 360 degrees in a circle, and that 24 divides into 360, is not a miracle of nature. That said, much of the significance we attach to numbers stems from our observation of natural fact: the number of fingers on each hand; the number of days and nights that pass between full moons; the number of planets visible to the naked eye.

This book is a tribute to the charisma of numbers. There are numbers from nature, mathematics, science, religion, mythology, superstition, art, history, technology... In an effort to apply some structure to this mind-boggling subject, I have included every whole number from 0 to 100 (plus a few notable imperfect numbers), and then picked out a selection of larger numbers that should either be familiar to everyone, or relate to something that is familiar. If I've missed out your favorite number, I apologize. This is not a definitive list. How could it be? The choice is infinite.

Tim Glynne-Jones

Is zero a number? If you're one of those people who insist that white is not a color, you probably think not. After all, it's neither positive nor negative. It is simply nothing, so how can it exist as a number? Well, as the saying goes, if you can put money on it in Vegas, it exists. But, in comparison with 1 to 9, it is a very recent discovery.

66 *God made everything out of nothing, but the nothingness shows through.* **99**

Paul Valéry, French poet and philosopher

THE ANCIENT GREEKS did not recognize 0 as a number. The people who mastered geometry and calculated pi were baffled by 0. As were the Romans. In India, where the number system we use today originated, the Hindus had some concept of it as a part of bigger numbers like 10 and 100, where it serves as a place-holder to show that the figure 1 represents 10s or 100s rather than units. They wrote it as a dot, which may have been enlarged to a ring, to give us the now familiar 0. An inscription dated 876AD shows use of a 0 as we would recognize it today.

Y E A R   Z E R O

MUCH of the evidence of ancient counting systems is gleaned from calendars. There is no year 0 in our Gregorian calendar, but for the Mayans, who flourished in South America in the first millennium AD, time began at "day 0," a day that has been calculated to correspond with August 11 3114BC. The **Mayans** had various calendars for different purposes (see **20**), one of which is called the **Long Count system**, by which they plotted significant dates over a great number of years. Beginning with 0 and then counting every day numerically, this relied on the use of 0 in a way that other counting systems of that era did not. And, unlike anybody else, they had a special symbol, a shell, for 0.

Fast-forward to 1975 and the Year Zero takes on a far more sinister significance. That year, when the **Khmer Rouge**, led by Pol Pot, seized control of Cambodia, they changed the calendar to Year Zero and erased all that had gone before. Anyone who was perceived to be a threat to the regime was executed. You could be killed for simply wearing glasses, as that was regarded as a sign of being an intellectual, and intellectuals were a threat. By the time Pol Pot's killing spree was brought to an end in 1979, an estimated 1 to 2 million people had been killed.

■ **The term "love,"** meaning 0, in tennis is derived from the French "l'oeuf," meaning "the egg"—an egg looking not unlike a 0. The same thinking gave rise to the use of "duck" in cricket for a batsman's score of 0. Scratch is the golfing term for 0, as in "scratch out," meaning to erase. A "scratch golfer" plays off a 0 handicap. Nil is a term that is rarely used outside the field of sport (one exception being the medical phrase "nil by mouth," meaning "not to be swallowed"). Nil is a simple abbreviation of "nihil," the Latin word for nothing.

❏ The most universal word for 0 today is zero. Like nil, it originated in Italy, thanks to the legendary mathematician Leonardo Fibonacci (see **1.618**). He took the Arabic word "sifr" (meaning empty) and gave it an Italian flourish, "zefiro," which was later abbreviated to "zero." It also gave us the word *zephyr*, for a faint, almost non-existent wind.

Heroes and zeros

Sport would be at a loss without 0. It's the point at which all games begin, and there are numerous ways of saying it:

● Nil ● Nought ● Zero ● Nothing ● Zip ● Zilch ● Nix (from the German *nichts*) ● Love ● Duck ● Scratch

❏ **GROUND ZERO** means the centre of an explosion or other disaster. For example, Ground Zero in Hiroshima, Japan, is the point above which the Atom Bomb exploded in 1945. And Ground Zero in New York is the site where the Twin Towers stood before 9/11. It is also the name given jokingly to the central plaza at the Pentagon, HQ of the US Defense Department, because it was considered the most likely target for attack during the Cold War.

✳ **The Mitsubishi Zero** (A6M) was an extremely potent Japanese fighter plane of WWII. It played a key role in the attack on Pearl Harbor, being designed for launch from aircraft carriers but still quick and agile enough to outmaneuver the US land-based fighters. It entered service in 1940 and it took most of the war for Allied air forces to come up with their own planes and tactics to counteract it. It took the name Zero from its designation as Navy Type 0 Carrier Fighter.

❝If everything on television is, without exception, part of a low-calorie (or even no-calorie) diet, then what good is it complaining about the adverts? By their worthlessness, they at least help to make the programmes around them seem of a higher level.**❞**

Jean Baudrillard

❏ 0 is the number of points traditionally scored by Norway at the Eurovision Song Contest after the votes have been counted from all over Europe. By 2006, the "nul points" order of merit stood as follows:

NORWAY (4)	GERMANY (2)
SWITZERLAND (3)	ICELAND (1)
FINLAND (3)	ITALY (1)
SPAIN (3)	LITHUANIA (1)
AUSTRIA (3)	MONACO (1)
NETHERLANDS (2)	YUGOSLAVIA (1)
BELGIUM (2)	SWEDEN (1)
TURKEY (2)	UK (1)
PORTUGAL (2)	LUXEMBOURG (1)

The years 2000–9 were dubbed the "noughties" because they make up the decade of 0s (noughts). Whether they were any naughtier than, say, the '60s, is open to question.

✳ Morroccan soccer player Hicham Zerouali, who played for Aberdeen in Scotland, was nicknamed "Zero" by the fans and wore the number 0 on his shirt.

✳ The city of Pontianak in Indonesia is located precisely on the Equator, at 0°0' N, 109°20' E.

✳ 0–100km/h (0–62mph) is the standard way of measuring a car's acceleration, the metric measure having replaced 0–60mph.

Absolute zero (−273.15°C) is the point at which the molecules of all substances have no energy, i.e. they freeze. All of them!

N O T H I N G   R E A L L Y   M A T T E R S

O has even spawned its own philosophy. Nihilism is the belief that nothing has any value, purpose or meaning. The term was coined by Russian author Ivan Turgenev in his 1862 novel *Fathers and Sons*, and it was the banner of a cultural movement that was said to have undermined the moral fabric of Russia and beyond. It spilled over into art and literature, becoming the central theme in the work of philosopher Friedrich Nietzsche (above), and influencing many other

philosophers such as Jean-Paul Sartre and Albert Camus, though they weren't nihilists themselves. Few own up to being a nihilist, but people who could be described as such include:

ADOLF HITLER
THE DADAIST "ANTI-ART"
MOVEMENT
SEMIOLOGIST JEAN BAUDRILLARD

JOHNNY ROTTEN
MARILYN MANSON
THE THREE NIHILISTS IN
THE BIG LEBOWSKI

1

Once upon a time there was but one number. The number one. And one is the most commonly used number in the world today. It's everywhere.

Good ones

HOLE IN ONE
NUMBER ONE
AT ONE
ONE OF A KIND
THE CHOSEN ONE
THE ONE AND ONLY
ONE-OFF

Bad ones

ONE-HIT WONDER
ONE-HORSE TOWN
ONE-TRICK PONY

Other ones

ONE-MAN BAND
ONE-WAY STREET
ONE FLEW OVER THE CUCKOO'S NEST

■ If a zero is a nobody, number one is the opposite extreme: the best, a winner, a leader, a favorite etc. It is also used to mean oneself, especially if you happen to be Queen of England: "One is not amused." But one is a lonely number and the Chinese believe it to be unlucky.

✳

Mono, from the Greek "monos," also refers to one.

Monochrome—one color

Monotheism—belief in one god

Monomania—an obsession with one thing

❝One is the loneliest number that you'll ever do
One is the loneliest number, worse than two**❞**

"One" by Harry Nilsen

❏ **1** is traditionally the number worn by the goalkeeper in soccer. Shirt numbers were first worn in the English league in 1928, with players numbered 1 to 11. The idea of squad numbers was introduced at the 1954 World Cup, and in 1978 Argentina stretched this further by numbering their World Cup team alphabetically. This meant Norberto Alonso, a midfielder, wore the number 1 shirt. Squad numbers at club level were also pioneered in England, in 1993, and the system remains in force today.

AS ONE

One is expressed in many different ways. The words *lonely*, *lonesome* and *loner* all stem from "alone," which is a shortening of "all one." *Solo*, a performance by one instrument, comes from the Latin *solus* meaning alone, as does *sole*, *solitary*, and *solitaire*—a game for one or a single gem set on its own. **Unit**, a single thing, gives us *unity*, *unite*, *unison*, *uniform*, *unique*, *unisex* and, of course, *united*.

✳

1 IS THE atomic number of hydrogen, which means there is only one proton (positively charged particle) in each hydrogen atom. This puts hydrogen at the top of the Periodic Table, which lays out all the known elements—of which there are currently 117 confirmed—in order of their atomic number. Hydrogen is reckoned to make up about three-quarters of the mass in the universe.

✳

Aces high The word "ace" comes from the medieval French, who used the word "as" for the one on a dice. Through its dual use in playing cards, it came to represent high scoring, as in the flying aces of the First World War, who scored a high number of "kills." Its use for an unreturnable serve in tennis stems back to the sense of one, as simply one shot played, one point scored.

ONE FOR THE MONEY

WHEN YOU STUDY a set of data, you might expect to find the numbers 1–9 appearing in roughly equal measure as the first digit, i.e., 11.1 per cent (1 in 9) each. However, an American physicist called Dr Frank Benford discovered that this is not the case. In fact, 1 appears as the first digit in almost a third of all cases (30.1 per cent). This probability decreases as you go up to 9, which only appears as the first digit 4.6 per cent of the time. By contrast, people who concoct fraudulent data tend to start their made-up numbers with 6 most commonly. These findings have inspired investigators to apply Benford's Law when checking for fraud. So if you're going to fiddle your tax return, throw in a few more 1s.

Research has also found that the number 1 puts ideas into people's heads. In a line-up, police omit numbering anybody 1, because it has been shown to influence a witness's choice.

■ In mathematics, 1 is the only number other than 0 whose square is the same as itself: 1 x 1 = 1. And here's an interesting set of sums involving 1:

$$1 \times 1 = 1$$
$$11 \times 11 = 121$$
$$111 \times 111 = 12,321$$
$$1111 \times 1111 = 1,234,321$$
$$11111 \times 11111 = 123,454,321$$

FAMOUS FIRSTS

First Lady (Martha Dandridge Custis Washington was the first)

"First Cut is the Deepest"

First Love, Last Rites

First past the post

First Among Equals

Top 10 One-Hit Wonders in the USA as compiled by American cable network VHI in 2002: **10. "Ninety-Nine Red Balloons" Nena** 9. "Rico Suave" Gerardo **8. "Take On Me" a-ha** 7. "Ice Ice Baby" Vanilla Ice **6. "Who Let the Dogs Out?" Baha Men** 5. "Mickey" Toni Basil **4. "I'm Too Sexy" Right Said Fred** 3. "Come On, Eileen" Dexy's Midnight Runners **2. "Tainted Love" Soft Cell** 1. "Macarena" Los Del Rio.

1.4142

As PROVEN BY **Pythagoras**, the celebrated Greek mathematician, if you have a right-angled triangle with two sides of 1 unit in length, the hypotenuse (the long side) will be $\sqrt{(1^2+1^2)} = \sqrt{(1+1)} = \sqrt{2} = 1.4142$. This is known as **Pythagoras' Constant** and can be used to determine the diagonal of a square.

Pythagoras' therorem also enabled a simple method for architects and builders to create right-angles. The Egyptians, for example, used a rope with knots at regular intervals forming 12 equal segments. This rope was then pegged out to form a triangle with sides of 3, 4, and 5 segments. The angle opposite the 5-segment side was then known to be a right-angle, since $5^2 = 3^2 + 4^2$.

Pythagoras' Constant
1.4142... × LENGTH OF SIDE

However, $\sqrt{2}$ is what's known as an **irrational number**, something in which Pythagoras refused to believe. An irrational number is one which cannot be expressed as a fraction, e.g., x/y where x and y are whole numbers. It was one of his students who, having tried to express $\sqrt{2}$ as a fraction, realized it was impossible and put forward the notion of irrational numbers. As legend has it, he was drowned on Pythagoras's orders for his audacity.

1.618

Phi—The Golden Number

Here's a question for you. What do the following have in common?

THE GREAT PYRAMIDS OF EGYPT
THE PARTHENON
NOTRE DAME CATHEDRAL
A SUNFLOWER
THE LAST SUPPER BY LEONARDO DA VINCI
A STRADIVARIUS VIOLIN
THE HUMAN BODY

All of these things share proportions that equate to 1.618... plus a load more decimal places—also known as phi, the Golden Number, the Golden Section and the Divine Proportion. The more you look, the more you find its influence. It applies in geometry, mathematics, nature and art, and it may just govern life as we know it.

Fibonacci and the Sound of Phi

Modern studies into the Golden Number have found that it has an effect on sound, and therefore can be applied to create superior acoustics in recording studios. Antonio Stradivari, the 17th-century violin-maker, would not have been aware of these studies, but he applied the Divine Proportion in the design of his instruments and the sound quality he achieved is second to none.

What "Stradivarius" would have known is that in any musical scale, there is a harmonious relationship between the 1st, the 3rd, the 5th and the 8th (octave), numbers which by then had been intrinsically linked with the Golden Number by a 12th-century Italian mathematician called Leonardo **Fibonacci**. (See p.18.)

GEOMETRY AND ARCHITECTURE

Draw a line. Now divide that line into two segments, so that the ratio of the small segment to the large segment is the same as the ratio of the large segment to the whole line.

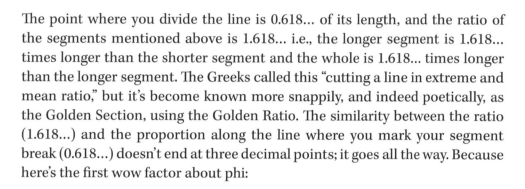

The point where you divide the line is 0.618... of its length, and the ratio of the segments mentioned above is 1.618... i.e., the longer segment is 1.618... times longer than the shorter segment and the whole is 1.618... times longer than the longer segment. The Greeks called this "cutting a line in extreme and mean ratio," but it's become known more snappily, and indeed poetically, as the Golden Section, using the Golden Ratio. The similarity between the ratio (1.618...) and the proportion along the line where you mark your segment break (0.618...) doesn't end at three decimal points; it goes all the way. Because here's the first wow factor about phi:

$$1/\text{phi} = \text{phi} - 1$$

You won't find that with any other number. The mathematicians among you will deduce from this another amazing equation:

$$\text{phi}^2 = \text{phi} + 1$$

Try it: 1.618... x 1.618... = 2.618... exactly.

★

The Ancient Egyptians and Greeks didn't need calculators that gave them phi to countless decimal places in order to use it. Their mathematicians worked out that the Golden Section can be derived by way of simple geometry, and hence applied on any scale they desired—even to a great pyramid.

Here's one way of doing it. Draw an equilateral triangle inside a circle so that the three corners touch the circle. Now draw a line that joins the midpoints of two sides of the triangle and extend the line to meet the circle. The distance between the midpoints = the distance from midpoint to circle times phi.

Phi governs the relationship between circles and other regular

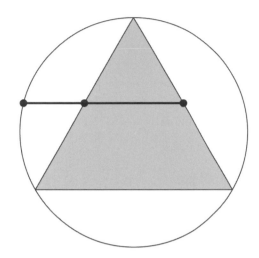

geometric shapes in a similar way, and this informed the ancient architects who were looking for perfect proportion in their buildings. Anyone who has visited the pyramids of Egypt or the Parthenon in Athens will agree that they were on to something.

Further maths

Leonardo Fibonacci wrote his name into history while making a study of rabbits. He wanted to work out how quickly their population increased if you started with two infants of the opposite sex. He plotted a table of population growth based on the pair mating at one month old, then giving birth a month later to another mating pair, which followed the same sequence, and so on. If you try this yourself, start with 0 and write down the number of pairs of rabbits at the end of each month (we're assuming no fatalities here). You will get a series of numbers that begins: 0 1 1 2 3 5 8 13 21 34 55 89... This is called the **Fibonacci Series**, and it goes on forever, the simple formula being that each number after the first 1 is the sum of the two previous numbers. A closer look at the relationship between the numbers in

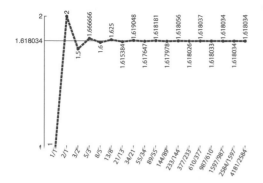

the Fibonacci Series reveals that, as you go up the scale, the ratio of one number to the next moves closer and closer to the Golden Number. So the Fibonacci Series is closely related to phi, the Golden Number, and thus takes its influence beyond the man-made world of mathematics and geometry.

ART

4,000 years after the Egyptians were sizing up the Great Pyramid of Giza, the artists and architects of the Renaissance period set great store by phi. They used it in their paintings and buildings, from *The Last Supper* to Notre Dame. It was identified in the proportions of the human face and body, as well as in other aspects of nature. No wonder they called it the Divine Proportion, for its appearance in so many aspects of life certainly must have hinted at some superior power at work.

Nature

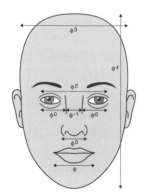

The Fibonacci Series is plain to see in the seeds, petals and branches of certain plants. The sunflower, for example, has its seeds arranged in spirals, whose number always conform to the series. Similar to the rabbits, many plants branch out in accordance with Fibonacci, first one branch, then two, then three, then five etc. It's actually a simple process of duplication, with each newcomer missing a go before commencing its own duplication process.

What Fibonacci would not have known is that the multiplication of plant and animal cells follows this sequence too, and this has been suggested as an explanation as to why so many objects in nature, e.g. the features on the human face and the spirals on a shell, fit the Divine Proportion. And the reason we find it such a pleasing, balanced proportion to behold may be nothing more complex than the fact that the human eye is built according to the same mathematical rule.

Once we've got used to the idea of our own existence, the concept of two is swift to follow. Two stands for sharing, co-operation, harmony. Conversely, it also means friction and opposition.

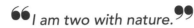

❝I am two with nature.❞

Woody Allen

OPPOSITES

This two-part symbol is called the Taijitu and lies at the heart of the Asian religion of Taoism. The two parts are called Yin and Yang, two universal opposites which must be in balance for the world to be at peace. Yin is the dark half, characterized as passive, shady, feminine, cold, mysterious, relating to the night. Yang, the light half, is active, bright, masculine, clear, hot, and associated with the sun. It has been widely adopted around the world as a symbol of harmony and balance, but actually in the Taoist belief Yin and Yang are constantly at war, and need to be balanced by a third party: man.

■ The Chinese place great importance on numbers according to their sound. The word for 2 (*uhr*) for example, sounds like "easy," and is therefore considered a good number. Put it with another good number, such as 8 (prosperity), and the portents are deemed very favorable. However, be careful how you use it. For example, four sounds like "death," so wearing the number 24 on your racing car would be seen as reckless.

■ **Double Trouble:** politicians love their buzzwords, and one of the most overused of the recent era has been "double whammy." The origin of this phrase rather reinforces suspicions about our respected leaders' bedtime reading: it is generally credited to a 1940s US comic strip called *Li'l Abner* by Al Capp. "Mudder Nature endowed me wit' eyes which can putrefy citizens t' th' spot! ... There is th' 'single whammy'..." the explanation goes. So a whammy is a spell cast by the evil eye, and a double whammy is one cast by both eyes: "...which I hopes I never hafta use." Al Capp probably got it from the word "wham," meaning "hit." "Wham" was first used in the *New York Times* in 1923. The last appearance of Wham was at Wembley Stadium in 1986.

Base 2

Imagine that our forebears had never invented more than two numerals, 0 and 1. Aside from being incalculably lazy, they would have left us with a counting system which goes 0, 1, 10, 11, 100, 101, 110... and by the time you wanted to write the year 2000 you'd have this figure: 11111010000. As with our decimal counting system, as soon as you run out of digits for one column, you start a new one. This, in essence, is the **binary system**, or base 2. And our forebears were tempted by it. Almost 3,000 years ago the Indian Pingala was dabbling with binary, and the ancient Chinese used it in their **hexagrams** (see 8). They were on to something, because today it forms the basis of the logic system used in computers. It was a British mathematician called George **Boole** who first put forward a system of logic based on binary, giving his name to Boolean Algebra. In 1937 this was applied by George **Stibitz** of Bell Labs, USA, in a system of electronic relays that formed the first electronic computer. And the rest, as they say, is nerdery...

**Translation of estate agents' doublespeak
from BBC Online:**

Bijou: Would suit contortionist with growth hormone deficiency.
Characterful: Old and falling down.
Charming: Pokey
Compact: See Bijou, then divide by two.
Four bedrooms: Three bedrooms and a cupboard.
In need of modernization: In need of demolition.
Mature garden: The local A to Z marks your garden as *Terra Incognita*.
Original features: Water tank still contains cholera bacterium.
Studio: You can wash the dishes, watch the telly and answer the front door
without getting up from the toilet.

SECONDS OUT

Why is a 60th of a minute called a second? It comes from the Latin phrase "pars minuta secunda," meaning second small part. This is also where the minute comes from. The phrase was used by the mathematician Ptolemy when dividing circles into smaller parts. One sixtieth of a circle he called "pars minuta prima" (first small part) and one sixtieth of that he called "pars minuta secunda." The terms were then applied to the divisions of the hour. Latin also gives us the sense of "second" as to assist or support, as in a boxer's corner men or somebody who supports a motion.

Examples of doubletalk from *The Quarterly Review of Doublespeak*:

• A doctor on the chart of a dead patient: "Patient failed to fulfill his wellness potential."

• Fleas—"hematophagous arthropod vectors."

• According to the US Army, they are "vertically deployed anti-personnel devices." Most other people know them as bombs.

• At McClellan Air Force base, California, civilian mechanics were placed on "non-duty, non-pay status." This means they were fired.

• Senator Orrin Hatch said that "capital punishment is our society's recognition of the sanctity of human life."

■ The Chinese have several Creation myths, but one of the most popular begins with an egg in which the giant Pangu grows for 18,000 years. When the egg hatches, the dark part of the egg drifts down to form the earth (Yin) and the light part floats up to form the heavens (Yang—see p.20). Pangu then places himself between them to keep them apart. After another 18,000 years, having grown at a rate of 10 feet a day, Pangu's work in keeping heaven and earth apart was done, and he died, his body parts going to make up the various features of the earth (wind, rocks, rivers, trees etc.). Mankind was made from the parasites on Pangu's body.

An object that is two-dimensional has length and width, but no depth.

★

Two is the only even prime number.

★

In the Inspector *cartoons that featured on* The Pink Panther Show *(1964 to 1971), Inspector Clouseau's sidekick is called Sergeant Deux-Deux (Two-Two).*

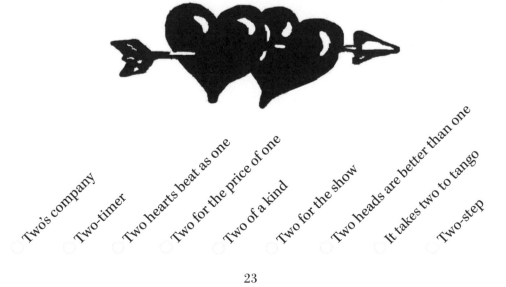

Two's company
Two-timer
Two hearts beat as one
Two for the price of one
Two of a kind
Two for the show
Two heads are better than one
It takes two to tango
Two-step

Never the Twain

MARK TWAIN was the pen name of Samuel Longhorne Clemens, the great 19th-century American, author. His most famous books featured the adventures of Tom Sawyer and Huckleberry Finn, but he was also a great wit, credited with such gems as: "I have never let my schooling interfere with my education," and "It is better to keep your mouth shut and appear stupid than to open it and remove all doubt."

Like his two great characters, Clemens' life was dominated by the Mississippi River, and working on the steamboats gave him the idea for his pen name. Safe water was two fathoms deep, and the boatmen would measure it with a marked line. In those days, "twain" was commonly used for two (as in "never the twain shall meet"), so the boatmen would cry out, "By the mark twain," on reaching safe water.

Twain has all but died out as an alternative word for 2, but plenty of others have survived: brace, couple, deuce, duo, pair, double.

Mixed doubles

Double top—the double 20 in darts, the topmost segment of the board.

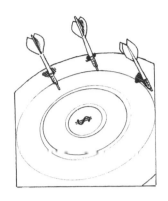

Double jeopardy—a legal term for the rule that prevents a suspect once acquitted from being tried again for the same crime, and a 1999 film starring Tommy Lee Jones.

Double Dutch—gobbledygook, and a skipping game using two ropes, brought to the world's attention by pop guru Malcolm McLaren during his "skipping" period, with his hit single of that name in 1983.

"Double double toil and trouble, fire burn and cauldron bubble"—the incantation spoken by the three witches in *Macbeth*, Act IV, Scene I.

Double or quits—a good way to recoup your money from a lost bet... or to lose even more.

Double time—money paid for working out of normal hours (twice the normal rate), and a musical term meaning doubling the tempo, often to get soldiers to march faster (at the double).

Doppelganger—the word for a lookalike, originally a ghostly double, from the German meaning "double-walker."

Folie a deux—a delusion shared by two people who live in close proximity. The classic case in Enoch and Ball's authoritative *Uncommon Psychiatric Syndromes* featured Margaret and her husband Michael, who were found to be sharing similar persecution complexes. They were convinced that certain persons were entering their house, spreading dust and fluff and "wearing down their shoes."

A favorite number among storytellers, three is a special number in both science and the arts. It stands for solidity, balance and completion.

❝ Now there are three steps to heaven
Just listen and you will plainly see **❞**

"Three Steps to Heaven"
by Eddie Cochrane, 1960

O UR ANCESTORS were very fond of the number 3 and it occurs repeatedly in scripture and mythology. Christianity has the Holy Trinity of the father, the son and the holy spirit. Islam has three holy cities, Mecca, Medina and Jerusalem. Yin and Yang (see **2**) were warring opposites, heaven and earth, that had to be balanced by a third party, man, and Taoism also has three deities called the Three Pure Ones. Brahma, Shiva and Vishnu are the trinity of Hindu gods, Buddha, Dharma and Sangha are the three "treasures" of Buddhism, and Norse mythology tells of three Norns, Urd, Verdandi and Skuld, who wove the tapestry of our fate, each person's life being a thread in the tapestry.

✳

Why is three such a popular number?

Here are three good reasons:

1 Storytellers are governed by the associations of certain numbers. One is a hero, a loner, not lending itself to dialogue and co-operation. Two is a romance, or a rivalry. And so, for a group of characters who act as one unit, with human interaction but without romance, the first and simplest choice is three. Of course, while a third party can provide the balance between two sparring factions, it can also put a dramatic spoke in the wheel—the classic *ménage à trois*.

2 When people try to perform a task in time with each other, e.g. lifting a wardrobe, they count to three before lifting in unison. This is because it takes a count of three to establish a rhythm, so that everyone heaves at the same moment. Running races are traditionally started with a three-part signal: on your marks, get set, go! Modern competition has dropped the "get set" part, going straight from "On your marks" to the "bang" of the gun, with a silent "get set" in between.

3 A man has 10 pairs of socks, 5 red, 5 blue, but he keeps them loose in his drawer. As he's dressing for dinner the light bulb in his room blows just as he's about to get his socks out, so he decides to take a few and put them on downstairs where he can see. How many socks would he have to take out of his drawer to ensure he had a matching pair? The answer is three, because if the first two are different colors, the third will definitely match one of them. If the first two are the same, that's the job done. This is a good explanation of why we say "third time lucky."

THREE OF A KIND

THREE LITTLE PIGS	THREE COINS IN A FOUNTAIN
THREE BILLY GOATS GRUFF	THREE MICHELIN STARS—THE
THREE BLIND MICE	HIGHEST ACCOLADE AWARDED
THE THREE BEARS	IN THE MICHELIN RESTAURANT
THE THREE MUSKETEERS	GUIDES
THREE WITCHES IN MACBETH	THE THREE VIRTUES—FAITH,
THREE WISE MEN	HOPE AND CHARITY
THREE TIMES A LADY	THE THREE GRACES—BEAUTY,
THREE STRIKES AND YOU'RE OUT	MIRTH, GOOD CHEER
THREE MEN IN A BOAT	THE THREE TENORS—LUCIANO
THE THREE STOOGES	PAVAROTTI, PLACIDO DOMINGO,
THE THREE AMIGOS	JOSÉ CARRERAS

TRI, TRI AND TRI AGAIN

Three is the smallest number of sides required to make a polygon: the triangle. And there are three types of triangle: **scalene** (all sides of different lengths), **isosceles** (two sides of the same length) and **equilateral** (all sides the same length).

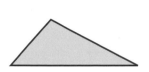
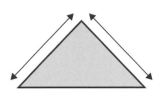
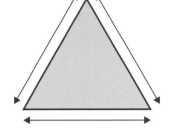

Three star

The predominance of threes in nature may also have influenced our interpretation of the number 3 as representing completion. Three dimensions; three elementary states: solid, liquid and gas; and astronomers have classified galaxies into three basic shapes: elliptical, spiral (and barred spiral), and irregular. They've even ascertained that elliptical galaxies are composed mostly of old stars, with little gas or dust. Spiral galaxies have plenty of gas and dust and their stars are a mixture of ages, while irregular galaxies are mostly composed of young stars. In other words, the older a galaxy gets, the more elliptical it becomes and the less dusty.

Tattoo you

■ In gang culture, numbers hold special significance, usually as a code to signify the gang name. 14, for example, is a common tattoo amongst the Nortenos, because N is the 14th letter of the alphabet. 18 is revered among far right groups because the numbers 1 and 8 correspond to A and H in the alphabet, the initials of Adolph Hitler (he of the Third Reich—third empire). But one of the most common tattoos of all is three dots in a triangle. This is seen among Hispanic gang members, and it stands for *Mi vida loca* ("My crazy life"), while for South-East Asians it denotes *To o can gica* ("I care for nothing").

In China, 3 ("sahn") has the same sound as "alive" and is therefore regarded as auspicious. Three-digit numbers are also favored for bringing luck.

3.14159 π

πd

r

d

Area=πr²

π IS THE GREEK letter for P, but it is so much more than that. It is an irrational number with an infinite number of decimal points, but generally speaking five or six are enough to use it extremely accurately.

π is the number used to calculate the area and circumference of a circle or ellipse. (It is the "p" of periphery that gave it its name). The circumference is π x **diameter**; area is π x **radius squared**.

The Greeks knew all this, although they didn't have the decimal system to write π the way we do. The closest they got was **Archimedes'** calculation that π was greater than 223/71 but less than 22/7, a very good approximation.

The quest to calculate π then moved east, where the renowned Chinese mathematician **Zu Chongzhi** narrowed it down to an approximation of 355/113.

This obsession among mathematicians continued up to the present day, during which time a Welshman named **William Jones** was the first to use the symbol π to denote *pi* in 1706.

A HUNGER FOR PI

On October 3 2006, **Akira Haraguchi** broke his own world record by memorising π to 100,000 decimal places. For most people, 10 decimal places is hard enough, so there are mnemonics designed to help, whereby you take the number of letters in each word. A common one is: "How I need a drink, alcoholic of course, after the heavy lectures involving quantum mechanics." This gives π to 15 decimal places. In 1996 **Mike Keith** wrote a short story called *Cadeic Cadenza*, in which the word lengths match the first 3,834 digits of π.

The number of legs on a table, chair or cow, four is a number of symmetry and stability, favored for groups and scientific endeavor.

> **❝** *Four hostile newspapers are more to be feared than a thousand bayonets.* **❞**
>
> Napoleon Bonaparte

THE EVENNESS and symmetry of 4 seems to have cost it in terms of playing a major role in storytelling. Being the first non-prime number and the square of two, it is too uniform to be used as a random number. Instead, it has served the purpose of dividing the natural world into neatly ordered sets: the four winds, the four cardinal points, North, South, East, and West, four seasons, four parts of the day: morning, afternoon, evening, night. The Greeks defined four elements: earth, air, fire, and water. It reflects a sense of balance, although in China they preferred to go a step further and divide the world into eight *(see* **8***)*. In fact, 4 is an unlucky number in the Far East, denoting death. And instead of the four elements defined by the Greeks, they had five: fire, wood, water, metal, and earth.

4/4 time is the most common rhythm used in music. It stands for four beats to the bar. Counting to four is a common method for bands starting a piece of music together, and punk bands of the 1970s such as the Ramones and the Clash made a feature of this, a statement of their anti-musicianship.

FOUR HORSEMEN

The most notable use of four in folklore is the biblical account of the Four Horsemen of the Apocalypse, who appear in the Book of Revelation, Chapter VI. These apparitions have been interpreted in several ways. The only one actually named is Death, so what the others represent is open to interpretation. Conquest, War and Famine are the usual assumption, although the first two may seem to be pursuing similar ends. Pestilence or Plague are popular alternatives, but it's difficult to deduce either of those from the King James version:

"And I saw, and behold a white horse: and he that sat on him had a bow; and a crown was given unto him: and he went forth conquering, and to conquer...

"And there went out another horse that was red: and power was given to him that sat thereon to take peace from the earth, and that they should kill one another: and there was given unto him a great sword...

"And I beheld, and lo a black horse; and he that sat on him had a pair of balances in his hand. And I heard a voice in the midst of the four beasts say, A measure of wheat for a penny, and three measures of barley for a penny; and see thou hurt not the oil and the wine...

"And I looked, and behold a pale horse: and his name that sat on him was Death, and Hell followed with him. And power was given unto them over the fourth part of the earth, to kill with sword, and with hunger, and with death, and with the beasts of the earth."

FAMOUS FOURS

The Fab Four—John, Paul, George and Ringo (aka The Beatles)
The four apostles—Matthew, Mark, Luke and John (the apostles)
The Four Tops—Motown group founded in Detroit, Michigan, in 1956, comprising Levi Stubbs, Renaldo "Obie" Benson, Lawrence Payton and Abdul "Duke" Fakir. Their most famous hit was "Reach Out I'll Be There" in 1966.
The Gang of Four—four Chinese Communist Party leaders, including Chairman Mao's wife Jiang Qing, together with Zhang Chunqiao, Yao Wenyuan and Wang Hongwen, who were arrested in 1976 after Mao's death and charged with instigating the Cultural Revolution in the 1960s which brought the country to the brink of civil war.

■ **Four-minute mile:** up until May 6 1954, the four-minute mile was perhaps the greatest unbroken barrier in athletics. That day, a 25-year-old British medical student named Roger Bannister ran the mile in 3 minutes 59.4 seconds. Fifty years later, the record for the mile had been cut to 3 minutes 43.13 seconds by the great Moroccan runner Hicham El Guerrouj.

❑ Plus-fours, the long shorts favored by golfers, were so called because they were four inches longer than knickerbockers.

■ "Four seasons in one day" is a phrase used in Australia and New Zealand to describe the extreme changes in weather that can take place in coastal areas in a very short space of time.

❑ The four mathematical rules: addition, subtraction, multiplication, division

■ Film Fours: Over 500 films have been made with "four" in the title.

❑ Journalism is the Fourth Estate. The other three are politics, the judiciary and the administration.

Four-letter word ・ Four-poster bed ・ Four-minute warning ・ Four fingers on each hand ・ Four suits of playing cards ・ Four movements in a classical symphony ・ Four seasons ・ Four chambers of the heart

FOURS, FORKS AND FAWKES

While pitchforks have two and tridents have three, dining forks generally have four tines. The man credited with introducing the fork to Britain was an Elizabethan named Thomas Coryat. A shameless social climber, Coryat was something of a laughing stock among the English aristocracy, and so set out to prove himself by traveling Europe, largely on foot, and writing about his adventures in a book entitled *Coryat's Crudities*. It's here that he describes the fork, which he saw being used as a dining implement in Italy (a country considerably more advanced than England at the time). He did the same for the umbrella.

But instead of gaining credit and respect for his part in introducing two such important symbols of Britishness, he was ridiculed by the Court, had his book plagiarized and received next to nothing for his efforts. So he set off again, this time reaching India, where he died in 1617.

Faring even worse than Coryat in England at that time was Guy Fawkes, the notorious member of the Gunpowder Plot of 1605 to blow up the Houses of Parliament.

Fawkes' punishment was to be hanged, drawn and quartered, meaning his body was cut into four parts for display at different locations. His head (a fifth part) was also displayed. All this occurred after he had been dragged through the streets on a hurdle (pretty painful), hanged but cut down before dead (very painful), and his entrails and genitals removed and burned before his eyes (deeply upsetting).

5

Being the number of digits on each hand and toes on each foot, it is little wonder that five holds so much significance for us.

66 *One and one is two, and two and two is four, and five will get you ten if you know how to work it.* **99**

Mae West

THE FIFTH NUMBER in the Fibonacci series and the number of planets visible to the naked eye— Mercury, Venus, Mars, Saturn, Jupiter—5 is also the number of the senses. It is the number of oceans and, according to the ancient Chinese and Indians, the number of elements (see **4**). It's the number of rings in the Olympic symbol as well as a popular number for pop bands and groups of adventurous children—although the fifth member of Enid Blyton's Famous Five was a dog called Timmy. We're used to clock faces being divided into measures of five seconds and minutes (see **12**), making five minutes a standard duration for a short break ("take five").

THE PENTAGON

Five is a key number for the HQ of the US Department of Defense, a.k.a. The Pentagon. Not only is it five-sided, but it consists of five concentric "rings" of corridors and there are five floors above ground (plus two below). This relatively low-level construction makes it all the more staggering that it is the world's most capacious office building, housing some 26,000 employees. There is no great significance in its pentagonal shape, although some theorists will tell you it's tied in with the Tudor Rose. In fact, it was built on land bordered by two roads that met at an angle that lent itself naturally to two sides of a pentagon.

Famous fives

THE JACKSON FIVE
FIVE GUYS NAMED MO
FIVE CHILDREN AND IT
HAWAII FIVE-O
THE FIFTH ELEMENT
PLEADING THE FIFTH

"Pushing through the market
square
So many mothers sighing
News had just come over
We'd got five years left to cry in"

"Five Years" by David Bowie

The pentagram, or five-pointed star, plays a significant part in many beliefs, from Christianity (where it is held by some to represent the five wounds of Christ) to Satanism, which uses an inverted pentagram (with two points up) to represent rebellion against Christianity. The Greek followers of Pythagoras also revered the pentagram, because, in its regular form, it contains the Golden Ratio (as described in 1.618...)

■ **On the count of 5**: 5 is a number that crops up repeatedly in music. Musical manuscript is written on a stave that consists of five lines; a perfect fifth is the interval between notes that provides the most "pleasing"-sounding harmony, and it is the interval used to tune the strings of a violin: G D A E. A classical quintet is usually a string quartet—cello, viola, two violins—with a fifth instrument, e.g. a piano or oboe.

❏ Fifth Columnists was a term that became popular during World War II for subversive elements who might be working for an enemy power while masquerading as loyal to their nation of residence. It originated in 1936 from a radio speech during the Spanish Civil War, in which nationalist General Emilio Mola, who had four columns of troops advancing on Madrid, referred to supporters within the city as his "fifth column."

V for Victory

BEETHOVEN'S 5th Symphony in C Minor is one of the most famous pieces of music ever written. And it played a significant part in World War II, though not, as Beethoven may have wished had he been alive, for the Germans. Instead, it helped to spread Winston Churchill's "V for Victory" campaign throughout Europe. So how did a German composer who died in 1827 come to symbolize V for victory? Is it because 5 is V in Roman numerals? No. It's to do with the opening four notes— three short Gs and a long E flat—that are so instantly recognizable as Beethoven's 5th. And in Morse code, three short dots and a long dash stand for the letter V.

The opening notes of Beethhoven's 5th were broadcast on the radio as part of Churchill's propaganda campaign, and the dot-dot-dot-dash rhythm caught on in various forms all over Europe and was tapped out as a statement of resistance to the German occupation.

■ Plato the Greek philosopher reckoned the classic elements (earth, air, fire and water) were composed of regular shaped solids, of which there are five. The first is a four-sided figure with triangular sides, the second a cube, the third an eight-sided figure with triangular sides (like two pyramids base to base), the fourth a 12-sided figure with pentagonal sides and the fifth a 20-sided figure with triangular sides. These are known as the Platonic solids.

The fifth taste: Up until 1987 it was the generally held belief that our tongues were built to detect four basic tastes: sweet, sour, salty and bitter. But researchers discovered a new type of taste receptor in the tongue which pick up a savory taste. This confirmed that there is a fifth taste, something that had been a long-held belief in Asia, and was known in Japan as *umami*.

Umami, meaning "savoriness" or "meatiness," was a term coined in 1908 by Professor Kikunae Ikeda of Tokyo Imperial University, who was trying to identify a taste which he said was common to meat, cheese, asparagus, and tomatoes.

F A I T H   I N   F I V E

The five pillars of Islam could best be described as the golden rules of the Muslim faith.

The first pillar is the belief that: "There is none worthy of worship except God and Muhammad is the messenger of God." This is the declaration of faith, or Shahadah.

The second pillar is the obligation to pray, and this is carried out five times a day facing in the direction of Mecca. The prayers are taken from the *Qu'ran* and must be spoken in Arabic.

The third pillar is the giving of money for the needy. Discounting such properties as a house or car, Muslims are obliged to give a fortieth of their capital each year for this purpose.

The fourth pillar is fasting, which takes place during the month of Ramadan. From dawn until sunset, Muslims go without food or drink. This is regarded as an exercise in self-purification and self-restraint.

The fifth pillar is the pilgrimage (*hajj*) to Mecca. Muslims who have the money and physical capability are expected to carry out the pilgrimage at least once in their life. Each year, two million find the wherewithal to visit Mecca, which, incidentally, is in Saudi Arabia.

Judaism is based on five books, known collectively as the Torah or Pentateuch. These are Genesis, Exodus, Leviticus, Numbers and Deuteronomy, the first five books of the Bible.

6

At first glance, six may appear an unassuming little number, but scratch the surface and you'll find six far more influential than you think.

66 Why, sometimes I've believed as many as six impossible things before breakfast. 99

Alice in Wonderland by Lewis Carroll

THE NUMBER 6 occurs frequently in sport, not least in American football where six points are awarded for a touchdown. You also get six for potting the pink in snooker and in cricket it is the number of balls per over. Six is the atomic number of carbon, the element on which all living things are based. It is the first number that's divisible by 1, 2, and 3 and it has other very special mathematical properties. Being half of the ubiquitous 12, the more you look, the more you realize that many things in life have been arranged into half-dozens.

Six feet under—the depth at which coffins are traditionally buried.

Six faces on dice

Six strings on a guitar: E A D G B E

Six dots is the highest number on a domino.

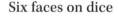

★

⊠ The Birmingham Six were six men sentenced to life imprisonment in Britain in 1975 for two pub bombings in Birmingham. Their convictions were overturned in 1991. The episode formed the basis of the 1993 movie *In the Name of the Father*, which was nominated for seven Oscars.

■ The largest group of species on earth consists of those with six legs, the insects. There are more than a million recorded species of insect, more than all the other animal groups put together.

❏ The world is commonly divided into six continents. However, this groups either North and South America or Europe and Asia as one. The other three are Africa, Australia (sometimes referred to as Australasia or Oceania), and Antarctica. Taking them all separately, there are, in fact, seven continents.

✳ *The sailor's navigation tool the "sextant" is so-called because its measuring scale covers 60˚, ⅙ of a circle.*

PERFECT SIX

6 has special qualities for mathematicians. It is both the sum and the product of its factors (the numbers that divide into it). In other words:

$$3 \times 2 \times 1 = 6$$
$$3 + 2 + 1 = 6$$

Try it with any other number. No, on second thoughts, don't: 6 is the only number that does this. It's not, though, the only number that is the sum of all its factors, but there aren't many. 28 is one (14 + 7 + 4 + 2 + 1), 496 is the next (248 + 124 + 62 + 31 + 16 + 8 + 4 + 2+ 1) and 8,128 is the next (4,064 + 2,032 + 1,016 + 508 + 254 + 127 + 64 + 32 + 16 + 8 + 4 + 2 + 1). These are called perfect numbers and these four are as far as the Greeks got in looking for them. But here are five more:

33,550,336

8,589,869,056

137,438,691,328

2,305,843,008,139,952,128

2,658,455,991,569,831,744,654,692,615,953,842,176

Six Count Swing, aka Boogie Woogie, is a rock'n'roll dance style of the 1950s.

★

A study has found that, while there are millions of sites on the internet, most of us visit just six on a regular basis.

Six of the best

Number 6—*the lead character, played by Patrick McGoohan, in the 1960s television series* The Prisoner.
Didier Six—*a member of the French soccer team that won the European Championship in 1984.*
The Kansas City Six—*a jazz band that featured Count Basie on piano.*
Six Degrees of Separation—*a play by American John Guare based on the theory that everyone in the world is connected to everyone else via a chain of no more than six acquaintances.*
Now We Are Six—*a collection of poems by A.A. Milne, the creator of* Winnie the Pooh.
Half A Sixpence—*classic 1967 film starring Tommy Steele and Julia Foster.*

> ## "Robbing people with a six gun I fought the law and the law won"
>
> "I Fought the Law"
> by the Bobby Fuller Four

A "six gun" or "six shooter" is Wild West slang for the classic revolver with six bullet chambers.

Circular Sixes

Six also has a special geometrical relationship with circles. If you take six identical circular coins and place them around another coin of the same size so that they touch it, they will all touch each other too. Whatever the size of the coins, 6 is the number that go round the outside. This can be proven using simple geometry. Place three coins so that they touch each other and the lines joining their centres form an equilateral triangle. The internal angle of an equilateral triangle is 60° and there are 360° in a circle. Therefore, six such triangles can be drawn from the centre of the first coin, giving six surrounding circles.

Six Quarks: A particle which has no smaller particles making it up is called a fundamental particle, and in particle physics it's these that are taken as being the basic elements of the universe. Physicists categorize these in three groups, one of which is "Quarks." And there are six types of quark: up, down, strange, charm, bottom and top; ridiculously simple names that conceal a branch of science designed to make your head implode. So let's leave it there. Except to say that the name "quark" was apparently invented by writer James Joyce, who used it to describe the noise a seagull makes.

The Six Wives of Henry VIII

CATHERINE OF ARAGON
(divorced)

ANNE BOLEYN *(executed)*

JANE SEYMOUR
(died two weeks after the birth of their son)

ANNE OF CLEVES *(divorced)*

CATHERINE HOWARD
(executed)

CATHERINE PARR *(survived)*

What a popular number seven is, cropping up again and again in religion and mythology. But what is the reason for this?

66 *I busted a mirror and got seven years' bad luck, but my lawyer thinks he can get me five.* **99**

Steven Wright, comedian

WE CAN ONLY speculate as to why 7 is writ so large in religion and mythology. Could it have something to do with the fact that we can see seven "heavenly bodies" of our solar system with the naked eye: the five planets mentioned in **5**, plus the sun and the moon?

Or could it be its sheer randomness that makes seven such a popular choice? Even numbers have symmetry, 1 has unity, 3 balance, 5 and 9 have mathematical uniformity. But 7 is much harder to pigeon-hole, and therefore lends itself well to representing an unspecific number of things.

Take the seven seas, for example. Any seafarer worth his salt would have known that there were more than seven seas. There's the North Sea, the Irish Sea, the Mediterranean Sea, the Caspian Sea, the Aegean, the Adriatic, the Black Sea, the Red Sea, the Dead Sea, the South China Sea... The word seven in this, and many other cases, may well have been used to mean "several."

The most common ladybird, *Coccinella septempunctata*, has seven spots—three on each wing and one at the nape of the neck. Ladybirds come in many varieties and their spots range from only two in some cases up to 24 in others.

The Seven-day Week

About 5,000 years ago, the Babylonians measured time by the appearance of the sun (1 day) and the cycle of the moon, which takes just over 29 days (roughly a month). But they wanted a measure that was shorter than this and, as 29 is a prime number, they decided the best they could do was divide it into four units of seven (28).

In English, the names of most days of the week were instigated by the Angles and Saxons, who replaced the Roman gods with their own equivalents and renamed the days accordingly.

Sunday—after the sun

Monday—after the moon

Tuesday—after Tiw, the Norse god of war. This replaced the Roman god of war, Mars, still evident in the *mardi*, *martes*, and *martedi* of the French, Spanish and Italians.

Wednesday—after the chief Norse god Woden. The Romans named it after the god Mercury (*mercredi* (fr), *miercoles* (sp), *mercoledi* (it)).

Thursday—after Thor, the Norse god of thunder, replacing the Roman equivalent Jupiter.

Friday—after Frigga, the Norse goddess of marriage and the hearth, replacing the Roman goddess of love, Venus.

Saturday—still named after the Roman god of time and the harvest, Saturn.

The seventh son of a seventh son is said to have the gift of second sight in many cultures. Depending on who you listen to, this will either make him a gifted doctor, a werewolf or both. So try to make sure your check-ups don't fall on a full moon.

In some cultures, including among certain tribes of Native Americans and the Kulin people of Australia, the year is divided into seven seasons.

"Give me a child until he is 7 and I will give you the man." So said Francis Xavier, a 16th-century Spanish missionary who spread Roman Catholicism into Asia. His belief was that the first seven years of life were when a person's character is formed.

The Austin 7 was one of the most popular cars ever produced. Built in Britain from 1922 to 1939, it sold close to 300,000. The Lotus 7 (later the Caterham 7) was a two-seater racing sports car, famously driven by No. 6 in *The Prisoner*.

★

American Roy Sullivan survived a record seven lightning strikes in 36 years as a park ranger in Virginia, USA. He was born at 7pm on February 7 1912. He died in 1983 from a self-inflicted gunshot wound.

The Seven Sisters is a constellation also known by the Greek name the Pleiades. In Greek mythology they were the daughters of Atlas and Pleione. The appearance of the Seven Sisters was the signal for the start of the Aztec "century."

THE SEVEN AGES OF MAN

"All the world's a stage,
And all the men and women merely players,
They have their exits and entrances,
And one man in his time plays many parts,
His acts being seven ages. At first the infant,
Mewling and puking in the nurse's arms.
Then the whining schoolboy, with his satchel
And shining morning face, creeping like a snail
Unwillingly to school. And then the lover,
Sighing like a furnace, with a woeful ballad
Made to mistress' eyebrow. Then a soldier.
Full of strange oaths, and bearded like the pard,
Jealous in honour, sudden and quick in quarrel,
Seeking the bubble reputation
Even in the cannon's mouth. And then the justice,
In fair round belly with good capon lined,
With eyes severe and beard of formal cut,
Full of wise saws and modern instances;
And so he plays his part. The sixth age shifts
Into the lean and slippered pantaloon,
With spectacles on nose and pouch on side,
His youthful hose, well saved, a world too wide,
For his shrunk shank; and his big manly voice,
Turning again towards the childish treble, pipes
And whistles in his sound. Last scene of all,
That ends this strange eventful history,
Is second childishness and mere oblivion,
Sans teeth, sans eyes, sans taste, sans everything."

As You Like It by William Shakespeare, Act II Sc VII

Seven digits is the most the average person can remember. This is why remembering a cellphone number is often tricky.

★

The Burj Al Arab in Dubai, designed in the shape of a sail, claimed to be the world's first seven-star hotel.

★

The drink 7Up began life as Bib-Label Lithiated Lemon-Lime Soda. It was invented by Charles Grigg in 1929.

★

Seven is the largest number of cylindrical objects (e.g. sticks) that can be tied securely in a bundle. The case of the six coins in 6 shows that any larger number will leave the middle cylinder loose.

★

☐ If you break a mirror, it's said to bring seven years' bad luck. You can erase the curse by burying the pieces or running them in a stream.

■ The Seven Years' War was described by Winston Churchill as the first real World War.

From 1756 to 1763, Great Britain, Prussia and Hanover took on the combined might of France, Austria, Russia, Sweden and Saxony in Europe and the colonies. Portugal later joined the British side and Spain joined the French.

Seventh Heaven

Followers of certain religious faiths will argue that the seven-day week was the creation of God. Certainly the number 7 crops up repeatedly in Judaism. According to Genesis, God created the world in seven days. And the first sentence of Genesis, written in Hebrew, is riddled with sevens. In English it reads, "In the beginning God created the heavens and the earth." In Hebrew, this comprises seven words and 28 letters, and these break down into further groups of seven. The Sabbath (Shabbat) is the seventh day of the week. There are seven holidays in the Jewish year, two of which, Passover and Sukkot, last seven days. The Menorah candle-holder has seven branches, three on each side and one in the middle. Likewise, the star of David has six points and a centre, said to represent God. The list goes on.

In both Judaism and Islam, Heaven is said to be formed of seven levels. This may relate to the seven "heavenly bodies" which held ancient man in such awe, and in some cases were believed to be levels to which the spirit traveled after death. Whatever the origin, seventh heaven is generally accepted as the height of bliss.

In Japan, seven holds religious significance too. For example, in Japanese Buddhism there are Seven Lucky Gods, and they believe that people are reincarnated seven times, and death should be followed by seven days of mourning. The Shinto 7-5-3 festival welcomes girls aged seven into womanhood.

The Seven Deadly Sins	The Seven Heavenly Virtues
LUST	CHASTITY
GLUTTONY	MODERATION
AVARICE	LIBERALITY
ENVY	CHARITY
WRATH	CHARITY
SLOTH	MEEKNESS
PRIDE	ZEAL
	HUMILITY

The Original Seven Wonders of the World:

The Great Pyramid of Giza

The Hanging Gardens of Babylon

The Temple of Artemis at Ephesus

The Statue of Zeus at Olympia

The Mausoleum of Mausolos at Halicarnassus

The Colossus of Rhodes

The Lighthouse of Alexandria

The Seven Summits are the highest mountains on each continent:

CARSTENSZ PYRAMID (4,884M)	OCEANIA
VINSON MASSIF (4,892M)	ANTARCTICA
MOUNT ELBRUS (5,642M)	EUROPE
KILIMANJARO (5,895M)	AFRICA
MOUNT MCKINLEY (6,194M)	NORTH AMERICA
ACONCAGUA (6,962M)	SOUTH AMERICA
EVEREST (8,848M)	ASIA

ASIA & THE MIDDLE EAST
by numbers

Asia and the Middle East is the biggest and by far the most densely populated continent on earth. It has the greatest range of altitudes of any continent: 30,369 feet.

Area
17,212,000sq.miles/44,579,000sq.km

Percentage of global land mass
30%

Population (approx)
4,000,000,000

Population density (per sq.km) 90

Highest point
Mt Everest: 29,028ft/8,848m

Lowest point
Dead Sea: 1,341ft/409m bsl

Longest river
The Yangtze River: 3,434 miles/ 5,530km

Highest recorded temperature
54°C/129°F (Tirat Tsvi, Israel, 1942)

Lowest recorded temperature
-68°C/-90°F (Oimekon, Russia,1933)

Eight is more of a mathematical number than a mystical one. It features very little in religion or mythology, but crops up repeatedly in the way we structure the world and its contents.

66 That woman speaks eight languages and can't say no in any of them. 99

Dorothy Parker

■ Spiders have eight legs as do scorpions, ticks and mites. These are all arachnids, of which there are around 70,000 species. An octopus, which also has eight legs (hence the name), is not an arachnid, it is a cephalopod (meaning "head foot," since there's very little in between). An octopus also has three hearts.

★

"Two and eight" is cockney rhyming slang for "state." "He's in a right two and eight."

★

Eight is common to the Imperial system of weights and measures. It's much easier to divide into smaller amounts than 10:

8 fluid ounces in a cup
8 pints in a gallon
8 gallons in a bushel
8 tablespoons in a gill
8 stone in a hundredweight
8 furlongs in a mile

❏ A musical scale consists of eight notes. Why not seven? Because a scale includes the octave (eighth note) of the first note. There are, of course, seven "natural" notes (A, B, C, D, E, F and G), but a total of 12 notes if you count sharps and flats.

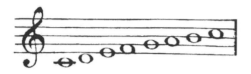

■ There are eight planets in the Solar System. They are, in order from the sun:

MERCURY
VENUS
EARTH
MARS
JUPITER
SATURN
URANUS
NEPTUNE

Pluto used to be classified as the ninth planet, but was demoted in 2006 to the status of dwarf planet under a new definition voted in by the International Astronomical Union. In fact, it had just been leapfrogged by the newly discovered Eris, another dwarf planet measured to be slightly larger than Pluto. The Solar System's third and smallest dwarf planet, Ceres, orbits in the asteroid belt between Mars and Jupiter.

■ The Number 8 in rugby is the player at the back of the scrum. It is one of the few positions in any sport denoted by shirt number. Other examples can be found in the traditional Oxford v Cambridge Boat Race that takes place every year on the River Thames in London. Each boat contains eight rowers, plus a cox. Positions one and eight are known as Bow and Stroke respectively, but the rest are known by their number from the bow. 8 is the number on the black ball in pool. Each player must pocket their own seven colored balls before they may pocket the "eight ball."

EIGHT IN ASIAN CULTURE

Just as Western religion and folklore seems obsessed with seven, in Asia eight is the special number. There are specific beliefs in the power of eight in the Far East. Its pronunciation in Chinese sounds the same as **"prosperity"** and so the number is associated with good fortune and potential. But eight also lies at the heart of Chinese philosophy, which centres around an eight-sided figure called the *Pa Kua* or **Bagua** ("eight symbols").

About 5,000 years ago, Chinese philosophers sought to organize what they saw as the fundamental elements of existence and to represent the constant state of change in the universe through the Bagua. In the centre of the diagram you see the **Taijitu**, the Yin and Yang symbol (see **2**). Around this is a series of **trigrams**, figures made up of three lines, some solid, some broken. The broken lines relate to Yin and the solid lines to Yang. Each trigram is a different combination of broken and unbroken lines—eight being the total number of ways that twos can be grouped in threes—and each has been given a set of attributes. There are **cardinal points**: North, South, East and West, plus North-East, South-East, South-West and North-West. Each cardinal point is associated with a **natural feature**: Heaven, Lake, Fire, Thunder, Wind, Water, Mountain, Earth; as well as **moods** and **personalities**; masculinity and femininity; eight members of a family: mother, father and three children of each sex; and so on. This was not just a random layout. South, for example, is where the Chinese saw the sun, so they placed Fire in that position. And they weren't the only ones to relate Mother and Earth.

The Bagua was thus seen as a **"map of life"** and given numerous applications, both practical and philosophical. It was used, for example, to plan the layout of cities, and today it applies in feng shui and martial arts. The eight trigrams were further combined to give **64 hexagrams** (groups of six lines), and these form the basis of the *I-Ching* ("Book of Changes") which lies at the heart of the three great Chinese philosophies, Taoism, Confucianism and Buddhism.

A list of eastern film titles featuring eight:

EIGHT TALES OF GOLD

VILLAGE OF THE EIGHT TOMBS

LITTLE PRINCE AND THE
EIGHT-HEADED DRAGON

THE EIGHT HILARIOUS GODS

EIGHT DIAGRAM POLE FIGHTER

EIGHT HOURS OF TERROR

THE INVINCIBLE EIGHT

HOUSEMAIDS FROM THE EIGHT PROVINCES

Some of these are Korean, some Japanese, some from China. The Eight Provinces ("paldo") is a term used for the whole of Korea, and in Japan eight also had the meaning of many.

■ Eight bells marked the end of the watch on board ship. From this it is thought that we derive the phrase "knock seven bells" out of someone, meaning to beat them to within an inch of their life. Knock eight bells out of them and you'd be up for murder.

❏ A V8 engine consists of two banks of four cylinders arranged together in a V-shape. This configuration is used to pack a lot of engine capacity (power) into a smaller space. French manufacturer De Dion is credited with developing the first mass-produced V8 engine in 1910, followed four years later by Cadillac, but Rolls-Royce appears to have been the first company to use a V8, in its Legalimit of 1904. However, only three were built, none of which survived.

> **"One over the eight," meaning "slightly drunk," originated as an army phrase, apparently based on the assumption that a man can drink eight pints (a gallon) of beer before getting intoxicated.**

9

The last single digit captured the human imagination almost as much as the heavenly seven. The fact that we spend nine months in the womb could have something to do with this, or perhaps it's just that nine times out of ten it's the first number that springs to mind.

SAYING "cats have nine lives" is generally taken as a random expression of the animal's ability to escape danger. However, it could have a more specific origin. The Egyptians deified cats, and also had nine gods. That said, in Arab countries (and Spain, which has absorbed considerable Arab influence in history), cats are said to have seven lives. So that scuppers that. Perhaps it has something to do with witchcraft and the number 9 being three 3s. In 1533 English author William Baldwin published a book called *Beware The Cat*, in which he wrote that "a witch may take her cat's body nine times."

■ "Possession is nine parts of the law" is an expression that applies in legal disputes over ownership. It is not a legal application as such; it just implies that it's much easier to keep hold of something you have than to take it from somebody else who claims ownership. In other words, possession is practically as good as legal right of ownership.

❏ The expression "as bent as a nine-bob note" is English slang for someone or something that is not bona fide.

"Bob" was the slang for the old British shilling, which came in ones, twos and 10s. Not 9s.

❏ "The whole nine yards" is a phrase said to stem from World War II, when fighter planes carried machine guns with 27-foot belts of ammunition. Give someone the whole nine yards and you'd be emptying your gun at them.

■ The US Supreme Court consists of nine judges.

A BIT OF MATHS

Nine is the square of three, which itself is a lucky number, but when you start playing mathematical games with nine, it throws up some interesting results.

The digits of all multiples of 9 add up to 9 or a multiple of 9. In fact, if you keep adding them until you only have one digit, it will always be 9. Let me demonstrate. Here's an easy one:

$$9 \times 9 = 81 \ (8 + 1 = 9)$$

And here's a more complicated one:

$$9 \times 137 = 1,233$$
$$(1 + 2 + 3 + 3 = 9)$$

And even more complicated still:

$$9 \times 3,641 = 32,769 \ (3 + 2 + 7 + 6 + 9 = 27; \ 2 + 7 = 9)$$

It works every time. Try it. Now try this. Take any three-digit number, invert it and subtract the smaller number form the larger number. The middle digit will always be 9. Look:

$$542 - 245 = 297$$
$$861 - 168 = 693$$
$$954 - 459 = 495$$

Magic, eh! And look at these multiples of 9:

$$1,089 \times 9 = 9,801$$
$$10,989 \times 9 = 98,901$$
$$109,989 \times 9 = 989,901$$
$$1,099,989 \times 9 = 9,899,901$$

...and so on. See what happens when you multiply those two magical numbers 9 and 7 together:

$$7 \times 9 = 63$$
$$77 \times 99 = 7,623$$
$$777 \times 999 = 776,223$$

The pattern continues:

$$7,777 \times 9,999 = 77,762,223$$

Nine dishes are traditionally served at a Chinese birthday banquet, the number 9 being associated with longevity.

M Y T H I C A L   N I N E S

Norse mythology is set in nine realms connected by a giant tree called Yggdrasil. At the top of the tree (Asgard) live the gods, and among the roots (Niflheim) lies the underworld, ruled by the goddess Hel. There are three realms in the nine levels which made up the Mayan underworld, Xibalba, with Metnal the ninth and most unappealing. And 9 was the number chosen by 14th-century Italian poet Dante in writing his epic about the afterlife, *The Divine Comedy*. In the poem he is guided through nine concentric circles of hell and nine spheres of heaven. In each case, the degree of evil and good increases at each level.

Dante was inspired by the Greek interpretation of the underworld, Hades, the centre of which was reached by crossing the river Styx, which circled Hades nine times.

The Greeks also believed in nine muses, the daughters of Zeus and Mnemosyne, who invented letters and their arrangement into art. Each represented a different "art" and were held dear by those who practiced those arts.

CALLIOPE—*epic poetry*	TERPSICHORE—*dance and*
CLIO—*history*	*choral song*
EUTERPE—*lyric poetry and song*	ERATO—*love poetry*
THALMIA—*comedy*	POLYHYMNIA—*sacred poetry*
MELPOMENE—*tragedy*	*and dance*
	URANIA—*astronomy*

Sudoku Nines

There are nine squares on each side of a Sudoku square. This Japanese number puzzle has become the global publishing phenomenon of the early 21st century, challenging the cryptic crossword as the daily brainteaser of choice. The rules are simple. You're given a 9 x 9 grid made up of nine 3 x 3 sub-grids, and a few numbers (at least 17, according to the latest theory) are already filled in. You have to fill in the rest of the grid so that the numbers 1 to 9 appear once in every row, every column and every sub-grid. It's a pastime that looks set to run and run: in 2005, Bertram Felgenhauer of the Department of Computer Science in Dresden, Germany, worked out that there are 6,670,903,752,021,072,936,960 different permutations on a sudoku grid.

■ "Dressed to the nines": nobody has explained the origin of this phrase, meaning dressed very glamorously.

❑ The cat 'o nine tails was a vicious whip used to instil discipline aboard ship. It comprised nine knotted strands which cut into the flesh.

■ Uzi 9mm—one of The Terminator's weapons of choice

❑ Federico Fellini's masterpiece, *8½*, was adapted for Broadway as the musical *Nine* in 1982.

■ The Beatles recorded a song called "Revolution 9" for the *White Album*, which consisted mainly of a voice intoning, "Number nine... number nine... number nine," specially designed to be spun round 1960s turntables.

The Nine Worthies are heroic figures from history, frequently referred to in literature:

HECTOR

ALEXANDER THE GREAT

GODFREY OF BOUILLON

JULIUS CAESAR

JOSHUA

DAVID

JUDAS MACCABAEUS

KING ARTHUR

CHARLEMAGNE

JVLIVS CÆSAR
(from the Naples bust)

10

The number of digits on both hands, ten became the natural base for the counting system we use today, which makes it the easiest number to work with in mathematics.

66*Grief makes one hour ten.***99**

Richard II, Act I Scene III,
by William Shakespeare

THE PRACTICAL USAGE of the number 10 may account for its relative scarcity in religion and mythology, although there is one notable example in Christianity:

THE 10 COMMANDMENTS

The one major exception is the Ten Commandments, God's instructions to the Israelites, as described in Exodus and Deuteronomy in the Bible. The Commandments were inscribed on two tablets by God and handed down to Moses on Mount Sinai. In essence, this is what they said:

1 YOU SHALL HAVE NO OTHER GODS BUT ME

2 YOU SHALL NOT WORSHIP FALSE IMAGES

3 YOU SHALL NOT TAKE GOD'S NAME IN VAIN

4 REMEMBER THE SABBATH DAY AND KEEP IT HOLY

5 HONOR YOUR FATHER AND MOTHER

6 YOU SHALL NOT MURDER

7 YOU SHALL NOT COMMIT ADULTERY

8 YOU SHALL NOT STEAL

9 YOU SHALL NOT BEAR FALSE WITNESS AGAINST YOUR NEIGHBOR

10 YOU SHALL NOT COVET YOUR NEIGHBOR'S POSSESSIONS

METRICATION AND DECIMALIZATION

The metric standard of weights and measures was a French invention, adopted after the Revolution of 1789 and later made widespread by **Napoleon Bonaparte**. Up until that point, a variety of weights and measures had been used, varying from one country to another. Metrication enabled goods to be traded internationally with a degree of consistency. By the end of the 19th century, most of Europe had converted to the metric system, and by the end of the 20th Century, all but a handful of countries in the world had followed suit, the USA being one notable exception. Interestingly, **Thomas Jefferson** had proposed a metric system in 1790, but Congress plumped for his second option, a development of the existing British system. The first units were the metre (hence "metric") and the gram. These standards are governed by the Système International d'Unités (International System of Units), from which we get the abbreviation **SI**.

The French also tried to introduce a metric clock, with 100 seconds in a minute and 100 minutes in an hour, but this was deemed a revolution too far and did not catch on.

A similar conversion has taken place in the world's currencies. **Decimalization**—the conversion to a base 10 currency system— was initiated by **Russia** in 1710, when the **ruble** was set at 100 kopecks. While the USA rejected the metric system of weights and measures, they did adopt a decimal system of currency, introducing the **dollar** in 1792. The French **franc** came in alongside the metrication process, but many countries held out until the mid-20th century, when the decline of the British Empire saw a number of Commonwealth countries decimalize. Britain itself did not decimalize until 1971, along with Ireland. The Japanese **yen**, established in 1871, is a decimal currency, as is the Chinese **yuan**, the Arabic **dirham,** and the Euro, introduced in 2002.

Number 10

No 10 Downing Street in London is the official residence of the British Prime Minister, and a popular tourist attraction. But who are the neighbors?

No 9 Downing Street—entrance to the Privy Council Office and the office of the Chief Whip

No 11 —home to the Chancellor of the Exchequer

No 12 —the Prime Minister's press office

The function of each building has changed over the years. For example, No. 11 is bigger than No. 10, and so there have been occasions where the Prime Minister and the Chancellor of the Exchequer have swapped houses because one had a bigger family than the other.

■ Decimate means to reduce something by one-tenth.

❏ Ten years is the maximum a US president may serve in office. He may only be elected for two terms (of four years each) but may accede to the position from vice president and hold it for two years.

■ 10 is the standard count in boxing after a knock-down. In 202 fights, the great Sugar Ray Robinson never took a 10 count. (10 is also the number you should count to when annoyed.)

❏ **Cloud 9 plus 1:** There are ten main types of cloud, divided into high, medium and low.

High clouds
Cirrus
Cirrocumulus
Cirrostratus

Medium clouds
Altocumulus
Altostratus
Nimbostratus

Low clouds
Stratocumulus
Stratus
Cumulus
Cumulonimbus

THE BILL OF RIGHTS

Drawn up by the First Federal Congress on September 25 1789,
the Bill contains ten amendments to the US Constitution

1 Congress shall make no law respecting an establishment of religion, or prohibiting the free exercise thereof; or abridging the freedom of speech, or of the press; or the right of the people peaceably to assemble, and to petition the government for a redress of grievances.

2 A well regulated militia, being necessary to the security of a free state, the right of the people to keep and bear arms, shall not be infringed.

3 No soldier shall, in time of peace, be quartered in any house, without the consent of the owner, nor in time of war, but in a manner to be prescribed by law.

4 The right of the people to be secure in their persons, houses, papers, and effects, against unreasonable searches and seizures, shall not be violated, and no warrants shall issue, but upon probable cause, supported by oath or affirmation, and particularly describing the place to be searched, and the persons or things to be seized.

5 No person shall be held to answer for a capital, or otherwise infamous crime, unless on a presentment or indictment of a grand jury, except in cases arising in the land or naval forces, or in the militia, when in actual service in time of war or public danger; nor shall any person be subject for the same offense to be twice put in jeopardy of life or limb; nor shall be compelled in any criminal case to be a witness against himself, nor be deprived of life, liberty, or property, without due process of law; nor shall private property be taken for public use, without just compensation.

6 In all criminal prosecutions, the accused shall enjoy the right to a speedy and public trial, by an impartial jury of the state and district wherein the crime shall have been committed, which district shall have been previously ascertained by law, and to be informed of the nature and cause of the accusation; to be confronted with the witnesses against him; to have compulsory process for obtaining witnesses in his favor, and to have the assistance of counsel for his defense.

7 In suits at common law, where the value in controversy shall exceed twenty dollars, the right of trial by jury shall be preserved, and no fact tried by a jury, shall be otherwise re-examined in any court of the United States, than according to the rules of the common law.

8 Excessive bail shall not be required, nor excessive fines imposed, nor cruel and unusual punishments inflicted.

9 The enumeration in the Constitution, of certain rights, shall not be construed to deny or disparage others retained by the people.

10 The powers not delegated to the United States by the Constitution, nor prohibited by it to the states, are reserved to the states respectively, or to the people.

EUROPE
by numbers

Europe, the birthplace of metric-ation (see p.59), is the second most densely populated continent on earth and its people speak a total of 45 different languages and dialects. They also consume over 60 per cent of the world's wine.

Area
3,837,000sq.miles/9,938,000sq.km

Percentage of global land mass
7%

Population
750,000,000

Population density (per sq.km)
75

Highest point
Mount Elbrus: 18,510ft/5,642m

Lowest point
Caspian Sea: 92ft/28m bsl

Longest river
Volga: 2,290 miles/3,700km

Highest recorded temperature
50°C/122°F (Seville, Spain, 1881)

Lowest recorded temperature
-55°C/-67°F (Ust'Shchugor, Russia)

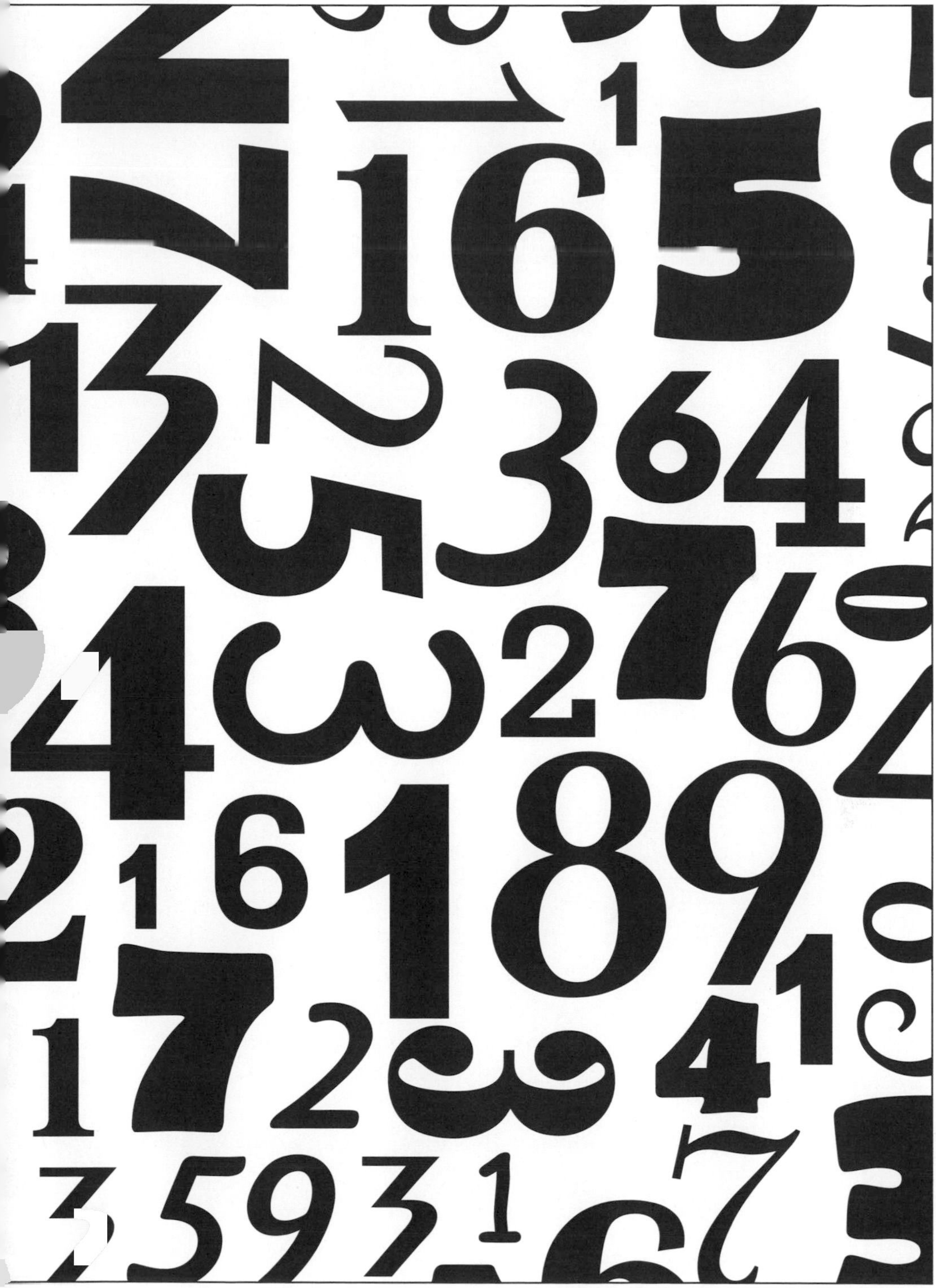

The fifth prime number, 11 nevertheless has a pleasing symmetry about it. It is simple to multiply by 1–9, and any power of 11 ends in a one. In many countries it is the age at which children start senior school; it marks one of the darkest days in American history; its place between 10 and 12 (the two most common counting bases) gives it double significance as "one more" or "one less," and it has special significance for fans of the cult movie *This Is Spinal Tap*.

Nigel: *"This is a top to a, you know, what we use on stage. But it's very very special because, if you can see... the numbers all go to 11. Look, right across the board, 11, 11, 11, 11."*
Marty: *"Most amps go up to ten."*
Nigel: *"Exactly."*
Marty: *"Does that mean it's louder? Is it any louder?"*
Nigel: *"Well, it's one louder, isn't it?"*

This Is Spinal Tap (1984)

In this "magic triangle," all sides add up to 11.

$$2$$
$$3 \quad 5$$
$$6 \quad 1 \quad 4$$

★

❏ September 11 2001 (9/11), American Airlines flight 11 was flown into the North Tower of the World Trade Center in New York. Was the number 11 perhaps a symbol of the twin towers?

■ The film *Ocean's Eleven*, remade in 2001 starring George Clooney, Brad Pitt, and Julia Roberts, was originally filmed in 1960 with Frank Sinatra, Dean Martin and Sammy Davis Jr, a.k.a. the Rat Pack. The Ocean is Danny Ocean, mastermind of a plan to rob three Las Vegas casinos, and the 11 are the members of his gang. A sequel, *Ocean's Twelve*, was made in 2004, followed by *Ocean's Thirteen* in 2007. The most recent film, *Ocean's Eight*, 2018, had a female cast.

The 11th Hour

The 11th hour of the 11th day of the 11th month marks the commemoration of the end of World War I, the armistice. The declaration was actually signed at 5am, aboard the private railway carriage of French Allied Supreme Commander Ferdinand Foch, and brought into force six hours later. In 1940, Adolf Hitler made a point of forcing the French to sign a peace deal aboard the same railway carriage.

Between 1914 and 1918, the Great War claimed close to 20 million lives. There follows a breakdown of estimated military deaths suffered country by country, to the nearest 1,000:

Africa (not South Africa) 10,000 • **Australia** 59,000 • **Austria-Hungary** 922,000 • **Belgium** 44,000 • **Britain** 659,000 • **Bulgaria** 88,000 • **Canada** 57,000 • **Caribbean** 1,000 • **France** 1,359,000 • **Germany** 1,600,000 • **Greece** 5,000 • **India** 43,000 • **Italy** 689,000 • **Japan** 300 • **Montenegro** 3,000 • **New Zealand** 16,000 • **Portugal** 7,000 • **Romania** 336,000 • **Russia** 1,700,000 • **Serbia** 45,000 • **South Africa** 7,000 • **Turkey** 250,000 • **USA** 59,000 •

If you leave something until the 11th hour, you're leaving it to the last minute.

★

Apollo 11 *was the spacecraft from which Neil Armstrong made the first moon landing in 1969.*

★

Elevenses is a quaint British expression for drinks taken at 11am.

66 *Well, what's there to talk about? Eleven men in here think he's guilty. No one had to think about it twice except you.* **99**

Juror #7 to #8 in the
1957 film *12 Angry Men.*

Number 8, played by Henry Fonda, having started out as the only member of the jury who doesn't think the accused is guilty, ends up winning all the other 11 jurors round to his point of view in dramatic style. The notion of trial by one's peers can be traced back to ancient Greece, but limiting the number on a jury to 12 was a Viking concept, later formalized by Henry II of England.

A   T W E L V E M O N T H

A twelvemonth is an old-fashioned expression for a year. "A twelvemonth and a day." The reason we have 12 months in a year is because that is the closest fit of lunar months (average 29.53 days) to solar years (365.24 days). Thus 12 became a significant number for the Sumerians, who divided the night into 12 stages, according to when certain stars appeared. To match this rough measurement, daylight was also divided into 12. The Romans counted the hours of the day from dawn until sunset, assigning certain hours to certain features of the day, and therefore the lengths of their hours varied according to the time of year. This basic system continued in Italy into the Middle Ages, when the first clocks as we know them were being invented, although they started counting 24-hour cycles from sunset. Elsewhere, in northern Europe, they preferred the Egyptian notion of two 12-hour periods rather than one of 24 hours, and designed their clocks accordingly. This also enabled greater detail on the clock faces, and prevented the need to chime so many times each hour after noon.

❏ Hitler's Third Reich lasted a mere 12 years rather than the thousand the Nazis envisaged.

FORCE 12 IS THE HIGHEST WIND SPEED ON THE BEAUFORT SCALE

0 CALM
1 LIGHT AIR
2 LIGHT BREEZE
3 GENTLE BREEZE
4 MODERATE BREEZE
5 FRESH BREEZE
6 STRONG BREEZE
7 NEAR GALE
8 GALE
9 SEVERE GALE
10 STORM
11 VIOLENT STORM
12 HURRICANE

■ *Twelve Monkeys* is a 1995 movie starring Bruce Willis and directed by Terry Gilliam. It was nominated for two Oscars, including Best Supporting Actor for Brad Pitt. During filming, Gilliam gave Willis a list of traits that he was not allowed to use. He called this "Willis acting clichés" and one of them was "the steely blue eyes look."

The 12 tribes of Israel are the descendants of the 12 sons of Jacob.

★

12 days of Christmas

★

The 12 apostles of Jesus

★

12 inches in a foot

★

"The Glorious 12th" (of August) marks the start of the grouse shooting season in Britain.

13

THE MOST superstitious of all numbers, 13 is even responsible for a recognized psychological condition, triskaidekaphobia—the fear of the number 13. According to research, this costs the US economy alone millions of dollars each year in absenteeism, cancellations and so on. Many tall buildings will skip from level 12 to 14, certain highways have no 13th exit and sports teams often leave out the number 13 on their shirt numbers.

Two notable sporting exceptions are the NBA basketball player Wilt Chamberlain, who always wore 13, claiming it was unlucky for his opponents, and legendary Miami Dolphins quarterback Dan Marino, who broke the record for yards passed, but never won a Super Bowl.

$$13 \times 13 = 169 \cdot 961 = 31 \times 31$$
$$13 - 1 - 3 = 9 = 3^2$$
$$13 + (1 \times 3) = 16 = 4^2$$

TEENAGE RAMPAGE

13 IS WHEN children tend to start behaving strangely, rebelling against their parents and authority. Perhaps that's why the number instils fear. It is certainly tied in with the onset of puberty. The culture of the "teenager" is a very modern phenomenon, arising in 1950s America, when rock 'n' roll and movies like *Rebel Without a Cause* starring James Dean tapped in to a growing desire among young people to speak out and break free from the constraints of the family unit. America witnessed a massive rise in juvenile delinquency during that decade. In Judaism it has always been seen as a coming of age for Jewish boys: the Bar Mitzvah. For Jewish girls (Bat Mitzvah), this happens at the age of 12.

SEAT 13

AIRLINES are particularly superstitious about the number 13, or at least their passengers are, and so they have no row 13 on board their planes (though why you should be more unlucky in that row than any other on board a plane is anyone's guess), and tend not to run a flight number 13. In the Far East, row 4 (which sounds like "death") is also omitted from some airlines, and there is no gate 4 or 44 at Inchon Airport in Seoul, Korea. Neither is there a gate 13. Flights with 7s and 11s, on the other hand, are considered lucky, and two US airlines run a flight 777 and 711 to Las Vegas. Perhaps this is why the 9/11 terrorists chose to hijack flights 77 and 11. In China, there's a lot of demand to get on board flight 88, the number standing for double prosperity. But statistically, flight number 191 is the one to avoid: two planes with that same flight number have crashed.

■ There are 13 major joints in the human body, three per limb plus the neck. This has provided some cultures with a method of counting, known as body counting. Counting fingers and toes is another form of body counting, and 5, 13 and 20 form the basis of the Mayan counting system (see **20**).

❑ 13 buns in a baker's dozen: in 13th-century England, bakers caught shortchanging customers could have a hand chopped off. To insure against this, they gave out 13 buns when asked for a dozen.

★

13 steps traditionally led up to the gallows.

Every 2.7 years there are 13 lunar cycles instead of 12, meaning one month will see two full moons. The second full moon is called a blue moon, hence the expression "once in a blue moon." This may have added to the bad reputation of 13, since the moon was associated  with lunacy and with the menstrual cycle—a severe case of PMT, in those days, being a short step away from being branded a witch.

14

FOURTEEN appears regularly in the Bible, whether it be rams, lambs, cubits, wives or whatever. But above all 14 is associated with love.

February 14th is Valentine's Day and it's the number of lines in a sonnet, the poetry form through which Shakespeare expressed so many romantic thoughts. The form of a sonnet is always the same: it uses the iambic pentameter, consists of three four-line "quatrains" with alternative rhyming patterns, and ends with a rhyming couplet. Shakespeare wrote 154 sonnets in total, this being the most famous:

"Shall I compare thee to a summer's day?
Thou art more lovely and more temperate.
Rough winds do shake the darling buds of May,
And summer's lease hath all too short a date.
Sometime too hot the eye of heaven shines,
And often is his gold complexion dimmed;
And every fair from fair sometime declines,
By chance, or nature's changing course untrimmed.
But thy eternal summer shall not fade
Nor lose possession of that fair thou ow'st;
Nor shall death brag thou wand'rest in his shade,
When in eternal lines to time thou grow'st,
So long as men can breathe or eyes can see,
So long lives this, and this gives life to thee."

14 STATIONS OF THE CROSS

The Stations of the Cross (or *Via Dolorosa*) is a Christian ritual that traces the key stages along the way Jesus took to his death. It began as a pilgrimage to Jerusalem and a walk along the path Jesus took towards his crucifixion at Calvary, pausing for thought at the 14 key landmarks along the way. For those who couldn't make the pilgrimage, depictions of the 14 stations were made and the ritual of prayer at each station became a part of the Christian faith, particularly around Easter and during Lent.

I JESUS IS CONDEMNED TO DEATH

II JESUS RECEIVES THE CROSS

III THE FIRST FALL

IV JESUS MEETS HIS MOTHER

V SIMON OF CYRENE CARRIES THE CROSS

VI VERONICA WIPES JESUS' FACE WITH HER VEIL

VII THE SECOND FALL

VIII JESUS MEETS THE WOMEN OF JERUSALEM

IX THE THIRD FALL

X JESUS IS STRIPPED OF HIS GARMENTS

XI CRUCIFIXION: JESUS IS NAILED TO THE CROSS

XII JESUS DIES ON THE CROSS

XIII JESUS' BODY IS REMOVED FROM THE CROSS (*PIETA*)

XIV JESUS IS LAID IN THE TOMB

■ Louis XIV of France, known as the Sun King, holds the record for the longest reign of any major European monarch—72 years. He ruled from 1643 to 1715, having come to the throne at the age of 4.

BLOODY VALENTINE

 The St Valentine's Day Massacre, which took place at the SMC Cartage Company on February 14 1929, was one of the most notorious incidents of the Prohibition era gang wars in Chicago. Though Al Capone was suspected of ordering the killing of seven members of rival Bugs Moran's gang, this was never proven, and no one was ever tried. Eye witnesses saw two policemen and two men in plain clothes leaving the scene. They assumed these were members of the Chicago police, calmly leaving the scene, having cleared up whatever problem had caused the gunfire, but in fact they must have been the killers, masquerading as policemen in order to get their victims to hand over their weapons and face the wall. In that position, the seven men were mercilessly gunned down.

> **66** *Inside outside, leave me alone*
> *Inside outside, nowhere is home*
> *Inside outside, where have I been*
> *Out of my brain on the five fifteen* **99**

"5:15" by The Who

The song above features in the 1979 British film *Quadrophenia*, the 5:15 being the train to Brighton on which Jimmy returns to find his hero working as a bell hop.

★

15 seconds is said to be the average amount of time an employer spends looking at a job applicant's CV.

★

"15 men on a dead man's chest, yohoho and a bottle of rum."

★

Philosophical 15s

"After fifteen minutes nobody looks at a rainbow."

Johann Wolfgang von Goethe, German playwright

"I have drunk since I was 15 and few things have given me more pleasure."

Ernest Hemingway, author

"Old is always 15 years from now."

Bill Cosby, comedian

"In the future, everyone will be world-famous for 15 minutes."

Andy Warhol, artist

"A revolution only lasts fifteen years, a period which coincides with the effectiveness of a generation."

Jose Ortega y Gasset, Spanish philosopher

THE FABULOUS 15TH CENTURY

The 15th century is dubbed the beginning of the Age of Discovery. It was hardly a peaceful time: the Hundred Years War rumbled on between England and France, including The Battle of Agincourt and the martyrdom of Joan of Arc; the Byzantine Empire fell into decline; the Spanish Inquisition gathered force; Beijing became the new capital of China; and the Wars of the Roses rocked the English monarchy. But amidst all this, the spirit of adventure and innovation burned strong. The New World was discovered and claimed by Spain. Chinese admiral Zheng embarked on seven voyages of discovery around the Indian Ocean, and a maritime link was opened between western Europe and India. The 15th century saw the first printing press and the first condom, while the Renaissance movement revolutionized art, music and science.

❑ **Fifteen** was the name of a charity restaurant in London set up by celebrity chef Jamie Oliver. Each year it took on 15 unemployed people from deprived backgrounds and trained them to work in the restaurant. The concept ran from 2002 to 2019, when the last restaurant closed.

■ 15 is a magic number for magic squares. It is the constant of a 3 x 3 magic square, i.e. every line in the square adds up to 15.

$$\begin{array}{ccc} 8 & 1 & 6 \\ 3 & 5 & 7 \\ 4 & 9 & 2 \end{array}$$

There are eight different ways of making a 3 x 3 magic square.

★

Fifteens are cakes originating from Northern Ireland. The recipe consists of 15 marshmallows, 15 digestive biscuits and 15 glacé cherries, plus ⅔ cup condensed milk and one cup dessicated coconut.

★

Fifteens is also a solo card game, in which you lay out 16 cards at a time and try to clear the deck by removing combinations that add up to 15.

16

" *She was only 16, only 16*
I loved her so
But she was too young to
 fall in love
And I was too young to know **"**

"Only 16" by Sam Cooke,
one of countless pop songs
about the magical age of 16

A ll around the world, sweet 16 is regarded as the first real step into adulthood. It is the age at which you may leave school and find a job, and in most countries it is the age of consent. In some places you can even drive a car at 16. Consequently, it has captured the imagination of song writers with a vengeance; everyone from Chuck Berry to Julie Andrews has sung about the magic of being 16.

Double 16 is the most
popular "checkout" in darts.
This is because 16 is next to 8,
so if you miss the double and
land in the 16, the next shot is
double 8, right next door.

Each player has 16 pieces in chess.

■ Magazines are generally printed in sections of 16 or 32 pages. As the sections are made by folding one large sheet of paper several times, they have to be multiples of 4, and at least four pages in size, as one fold gives four sides. The cover section of magazines will usually be just four pages because it is printed on a better-quality paper stock than the rest of the magazine, but the bigger the sections, the more economical, so 16- or 32-page sections will usually be used for the rest of the magazine.

★

16 ounces in a pound

★

16 is the number of named points on a compass.

★

Sixteen Hours

We spend an average 16 hours a day awake. A lion will be awake for about a quarter of that time. A study conducted by the UC Berkeley, California, found that the average American spends almost three of those hours watching TV, one hour 40 minutes driving a car, and less than 20 minutes per day doing physical exercise. Another study, by the US Bureau of Labor Statistics, recorded the following times spent on daily activities:

Eating and drinking	1 hour 15 minutes
Housework and cooking	1 hour 50 minutes
Shopping	40 minutes
Socializing and chatting	45 minutes
Physical exercise	17 and a half minutes

The M16 automatic rifle is the most widely used weapon of its caliber (5.56mm) in the world.

In China they developed a method of counting using the thumb to touch the tip and three joints of each finger. This enabled them to count to 16 on each hand.

In web design, colors are coded according to a base 16 numbering system. Taking the standard RGB (red, green, blue) values (0–255 for each), this systems converts them into base 16, using the numbers 0–9 and the letters a–f. **Example:** Yellow has an RGB value of 255, 255, 0. Its web color value is FFFF00 **Explanation:** 255 in base 16 = 1515 ((15 x 16) + (15 x 1)). 15 is represented by F.

APART FROM its widespread appearance in popular songs due to rock'n'roll singers' apparent obsession with teenage girls, 17 in itself is not a number that has lent itself easily to the development or culture of mankind. In fact, it has been shown to be the epitome of randomness. Studies have shown that if you ask people to pick a number between 0 and 20, 17 will come up more often than any other. Because of this it is called a "psychologically random" number, a number that is perceived as being more random than others. Because we see even numbers and multiples of 5 as having some pattern to them, numbers ending in 1, 3, 7 or 9 are often psychologically random,especially if they are prime numbers too.

*

TOO YOUNG TO DIE

In International Law, 17 is the maximum age at which you will be judged as a minor. Therefore, in countries where the death penalty applies, you may not be sentenced to death for crimes committed at or below the age of 17.

While 18 is almost universally regarded as the age of majority, most jurisdictions have elected to stagger the ages at which a young person attains certain legal rights. In the UK, for example, you can drive a car from the age of 17, even though you may not buy alcohol in a pub. Attaining the right to do both on your 18th birthday could represent a recipe for disaster.

In certain countries, particularly Ireland and Northern Ireland, as well as parts of Australia and several states of the USA, 17 is the age of consent. In most countries of the world, it is the case that at 17 you may legally be raising your own children, but are still not considered mature enough to vote.

In Italy 17 is regarded with fear. In Roman numerals it is XVII, and anagram of which is Vixi, which means I am dead. Tenuous as this may seem, it was enough for Renault to change the name of its R17 to R117 in Italy. And just as many buildings skip the 13th floor, Italian buildings skip the 17th, and some airliners have no row 17.

"When I was seventeen
It was a very good year
It was a very good year for
* small town girls*
And soft summer nights
We'd hide from the lights
On the village green
When I was seventeen"

"It Was a Very Good Year"
by Ervin Drake

The famous song to the left has been recorded in a wide variety of styles by a bewildering array of artists, from Ray Charles to William Shatner. The most successful version was sung by Frank Sinatra, for which he won a Grammy in 1966, but arguably the most memorable was the version sung by Homer Simpson:

"When I was seventeen
I drank some very good beers…"

■ **Japan's** famous haiku poems comprise 17 syllables over three lines, divided 5-7-5. The other rules are that it should contain some allusion to the seasons, and be formed of two parts, broken by a semi-colon. Here's an example:

"Numbers fill my brain;
Though I see the snow has come
I must write my book."

In Nordic culture, the 17th day of the year is considered the heart of winter.

❑ **Heaven 17** is an English pop band that rose to fame as part of the synthesizer pop movement of the early 1980s. They took their name from a fictional group in the Anthony Burgess novel *A Clockwork Orange*.

18

❝*I'm eighteen with a bullet*
Got my finger on the trigger, I'm gonna pull it.❞

"18 With A Bullet" by Pete Wingfield

The expression "with a bullet" comes from the Billboard chart, which put a bullet point next to records that had enjoyed particularly strong sales. Wingfield, whose song was indeed number 18 "with a bullet" in the Billboard Hot 100 in 1975, used the phrase as a shooting metaphor. The song peaked at 15.

18 HOLES

GOLF HASN'T always been played over 18 holes. In the early days at St Andrews, the famous Scottish course that claims to be the "home of golf," the course, carved out of the rugged coastal terrain, consisted of 11 holes. These were played twice in a round, making a total of 22. However, as the game developed, some of these holes were combined to make longer, tougher holes, and the total number was reduced to nine, giving a round of 18. (The bar at any golf club is colloquially known as the 19th hole.) In 2006, Welshman Bradley Dredge went round the Old Course at St Andrews in a record 64, six under par.

Are You 18?

18 is also the legal age for buying alcohol in most countries, with a few exceptions. In most Muslim countries, of course, alcohol is banned altogether. Other countries, such as Italy, have a minimum age for consumption but no minimum age for purchase. Here are some places that set a legal limit other than 18:

HAITI	6 (SCHOOL AGE)
SWITZERLAND	14
AUSTRIA, BELGIUM, CUBA, DENMARK, NETHERLANDS, PORTUGAL, FRANCE, GERMANY	16
GREECE, LUXEMBOURG	17
ICELAND, JAPAN, THAILAND	20
EGYPT, TUNISIA, USA	21

IN MOST democratic countries, 18 is the age at which a person may vote. In the USA, it is the age at which a person may bear arms, and the age at which they may be conscripted into the army. During the Vietnam War, the voting age in the USA was, in fact, 21 (as it used to be in many countries). The anomaly that a person could be drafted and die for their country, yet not have the right to vote, brought about a lowering of the voting age to 18 in 1971.

In Gematria (see **27**), 18 corresponds to the Hebrew letters *chet* and *yud*, which together make the Hebrew word for "life." The number 18 is therefore seen as significant among Jews, and at birthdays, weddings and Bar Mitzvahs it is customary to give money in multiples of 18, to symbolize the giving of life. Joseph Heller, author of *Catch-22*, originally intended to name his book *Catch-18* because of this significance.

*

February 18 1979: the day a 30-minute snow storm hit the Sahara in southern Algeria, the first time snow has been reported in the desert in recorded history.

18 characters is the limit for naming race-horses. That includes spaces between words.

19

THE LAST YEAR of teenage, 19 is special to many people on earth as the first two numbers of the year in which they were born. The 20th century was a century of major technological advance. Here are 19 things we take for granted today that were not around in the 19th century:

VACUUM CLEANER	1901	BALLPOINT PEN	1938
RADIO	1903	DIGITAL COMPUTER	1942
AEROPLANE	1903	MICROWAVE OVEN	1946
BRA	1913	CREDIT CARD	1950
STAINLESS STEEL	1916	HOVERCRAFT	1956
POP-UP TOASTER	1919	COMPACT DISC	1965
TELEVISION	1925	VIDEO RECORDER	1971
BUBBLE GUM	1930	WORLD WIDE WEB	1990
CANNED BEER	1935	VIAGRA	1998
NYLON	1935		

NINETEEN TO THE DOZEN

A mathematician will tell you that 19 to the dozen is $2.21331492 \times 10^{15}$, but that's not the point. The phrase is used to describe something that's going very fast, such as a runner's legs or somebody talking. Why it should be 19 is unclear. It may stem from the rate of fire during battle, or perhaps it was a measure of pulse rate. Before metrication, it would not have been unusual to count heartbeats every 12 seconds. The normal resting heartbeat of a healthy adult is anything between 60 and 100 beats per minute—12 to 20 every 12 seconds. Nineteen to the dozen has more of a ring to it than 20 to the dozen.

The phrase has become so detached from its origin that nobody is too sure what the original phrase is any more. "Ten to the dozen" is not uncommon, and you'll hear other numbers used to mean the same thing.

In 1642, in the build-up to the English Civil War, a list of proposals was put to King Charles I that would effectively transfer the power of the crown to Parliament. These were called the "Nineteen Propositions." Charles rejected the demands, saying that they threatened the ancient constitution of the kingdom and his lineage, and in doing so he removed the last barrier to open conflict.

★

19 is the number of the angels guarding hell in the Qu'ran.

★

"19th Nervous Breakdown" is a song by the Rolling Stones.

NI-NI-NI-NI-NINETEEN

"In 1965 Vietnam seemed like just another foreign war, but it wasn't.

It was different in many ways, as so were those that did the fighting.

In World War II the average age of the combat soldier was 26...

In Vietnam he was 19

In in-in-in-in-in-in Vietnam he was 19...

Ni-ni-ni-ni-19."

"19" by Paul Hardcastle, 1985

VIETNAM is often spoken of as America's costliest war, but in fact the casualties were relatively low even if the average age of combat troops was so young (see above). In terms of the percentage of the US population killed, Vietnam is on a par with the War of 1812.

WAR	DEATH TOLL	% OF POPULATION
War of Independence	4,435	0.15
War of 1812	2,260	0.03
Mexican War	13,283	0.06
Civil War	624,511	1.78
World War I	116,516	0.11
World War II	405,399	0.29
Korean War	35,516	0.02
Vietnam War	58,152	0.03
Gulf War	3,000+	

(Source: US Department of Defense)

THE ROARING TWENTIES was the ultimate decade of boom and bust. The 1920s began with the world eager to make up for the lost years of World War I and ended with the Wall Street Crash, the St Valentine's Day Massacre and the looming spectre of fascism. Jazz, penicillin, the mass-produced motor car, "talkies," television, trans-Atlantic flight and votes for women were offset against a background of Prohibition and political extremism. The roll call of legendary writers, artists and entertainers is vast: James Joyce, F. Scott Fitzgerald, Franz Kafka, George Bernard Shaw, T.S. Eliot, Ernest Hemingway, George Gershwin, Irving Berlin, Louis Armstrong, Cole Porter, Al Jolson, Charlie Chaplin, Greta Garbo, Joan Crawford, Buster Keaton, Harry Houdini... the list goes on and on, thanks largely to the media technology boom. With the exception of Russia, the world was letting its hair down, learning to live again after so much bloodshed. But before the decade was out, it would all turn sour and a new global scourge would be in place: the Great Depression.

■ The word "score," which is slang for 20 (as in three score years and ten—the traditional life expectancy of a person who survives infancy), most probably stems from the sense "to cut a mark." For every 20 in a count, a score would be made in a piece of wood or stone. From this, the word score took on the broader meaning of an overall count.

In France, 20 was used as a counting base, and this is still evident in the number *quatre-vingts* (four twenties), meaning 80.

Johann Sebastian Bach fathered 20 children.

20/20 is a term used for perfect eyesight. It comes from countries that use Imperial measures, 20/20 vision being normal vision at 20 feet.

COUNT TO 20

The Mayans had a base 20 counting system, subdivided into base 5. They used dots for units and a line for five, so 19 would be three lines with four dots. At 20, they put a dot in the next column to the left. They used this base 20 system for their Long Count calendar, which enabled them to log events far in the future. For some reason they used 18 as the base in the third column, probably because 20 x 18 (360) gave them a close approximation to the solar year. But after that it progressed in powers of 20: 1 day = kin; 20 kin = uinal (month); 18 uinal = tun (year); 20 tun = katun (7,200 days); 20 katun = baktun (144,000 days).

Taking the start of time as year 0,0,0,0,0, a date prior to their existence (see **0**), the Mayans counted their dates day by day from that date onwards. For example, 1966 solar years from year 0 would have been written as 4,19,14,12,1.

THE CURSE OF THE PRESIDENTS

American presidents were said to be afflicted with a "20-year curse." The president in office every 20 years from 1841–1961 died in office. Ronald Reagan, who became president in 1981, was shot in the head but survived.

1841	*William Henry Harrison* (1841)	died
1861	*Abraham Lincoln* (1861–65)	assassinated
1881	*James A Garfield* (1881)	assassinated
1901	*William McKinley* (1897–1901)	assassinated
1921	*Warren G Harding* (1921–23)	died
1941	*Franklin D Roosevelt* (1933–45)	died
1961	*John F Kennedy* (1961–63)	assassinated
1981	*Ronald Reagan* (1981–89)	survived attempt
2001	*George W Bush* (2001–09)	survived
2021	*Joe Biden* (2021–present)	TBC

SOUTH AMERICA
by numbers

South America is the longest continent from North to South and it boasts the world's longest mountain range, the Andes, which stretches for 4,500 miles.

Area
6,880,000sq.miles/17,819,000sq.km

Percentage of global land mass
12%

Population (approx)
390,000,000

Population density (per sq.km)
22

Highest point
Aconcagua: 22,835ft/6,962m

Lowest point
Bahia Blanca, Argentina: 138ft/42m

Longest river
The Amazon: 3,915 miles/6,300km

Highest recorded temperature
49°C/120°F (Rivadavia, Argentina, 1905)

Lowest recorded temperature
-33°C/-27°F (Sarmiento, Argentina, 1907)

Languages spoken
45

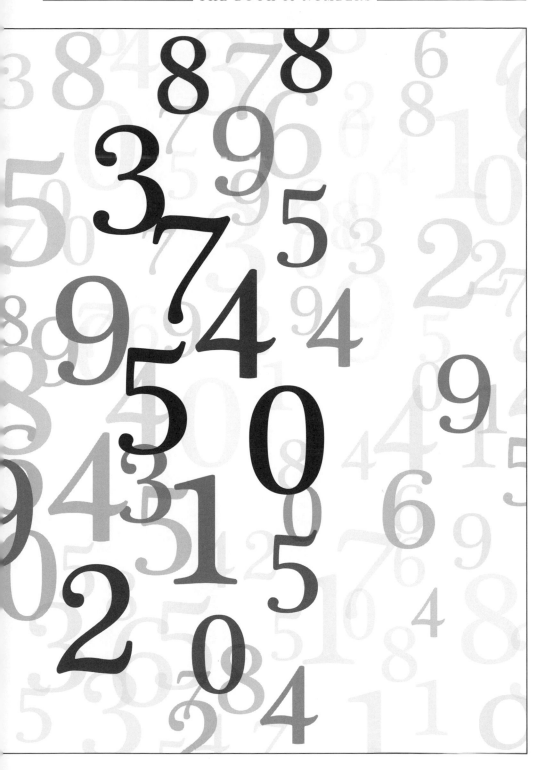

21

66 *My father told me all about the birds and the bees, the liar— I went steady with a woodpecker till I was twenty-one.* **99**

Bob Hope, comedian

TURNING 21 used to be regarded as the coming of age, until this was lowered to 18 in most countries. Remnants of this tradition still remain, for example in the USA where a person cannot buy alcohol until they are 21. Why 21 was chosen is unknown. Perhaps it's the superstitious significance of 21 being the product of the two numbers most associated with good luck, 7 and 3. Or it may once have been regarded as middle age, or indeed an age beyond which you were deemed to have survived the threats to health that kept average life expectancy to around that age.

❋

■ **A 21-gun salute** is fired in honor of royalty or heads of state. It is a mark of peaceful intentions, originating from the tradition amongst warriors to hold their weapons in a non-threatening way to signify no hostility. The advent of guns and cannons led to a seven-gun salute being fired aboard ships to signal this same passive intent— the firing of all your cannons rendering them temporarily ineffective because of the time required to reload. On land, this trebled to a 21-gun salute because gunpowder was not in such short supply. As the keeping of gunpowder improved, naval salutes were increased to 21 guns as well.

There are 21 dots on a die.

CATCH-22 was an expression invented by author **Joseph Heller** in his famous novel of the same name. It tells the story of **Yossarian**, a US bomber pilot during World War II, stationed on an island near Italy. Yossarian is constantly looking for ways to get out of flying missions, but is always thwarted by a catch, Catch-22. The following passage goes some way to explaining it:

66There was only one catch and that was Catch-22, that specified that a concern for one's own safety in the face of dangers that were real and immediate was the process of a rational mind. Orr was crazy and could be grounded. All he had to do was ask; and as soon as he did, he would no longer be crazy and would have to fly more missions. Orr would be crazy to fly more missions and sane if he didn't, but if he was sane, he had to fly them. Yossarian was moved very deeply by the absolute simplicity of the clause of Catch-22 and let out a respectful whistle.**99**

There are 22 players on the field in American Football, soccer and field hockey.

✳

❏ By the old English measuring system there are 22 yards in a chain. This is the length of a cricket pitch.

◼ The f-stop on a camera goes up to 22. The other stops are 2.8, 4, 5.6, 8, 11 and 16.

★

"Vingt-deux!" (22) is a coded warning in French that the police are coming.

22 bones make up the human skull.

23

ANOTHER PRIME NUMBER with no particular pattern to it—other than the fact that both its digits are primes and they add up to a prime—that has attracted believers in the paranormal. *23rdians* are a group of people who claim to see the number 23 everywhere they look and interpret this as a signal that they should take over the world.

Julius Caesar was stabbed 23 times, according to accounts of the time from writers such as Plutarch. Shakespeare embellishes this in the words of Octavius to "three and thirty" (Act V Scene I).

In 2007, actor Jim Carrey starred in *The Number 23*, a movie inspired by 23rdianism. Carrey himself even changed the name of his company to JC23, saying, "I've been obsessed with the number for years."

"The Lord is my shepherd; I shall not want.
He maketh me lie down in green pastures..."

Psalm 23

The 23-Day Cycle

Biorhythm theory states that men's physical and emotional characteristics follow a 23-day cycle. This was first put forward by Wilhelm Fleiss, a German physician and numerologist, who was a close friend of Sigmund Freud. Fleiss also found that women follow a 28-day cycle, although this was nothing new. In 1904, **Dr Horman Swoboda** produced what he claimed were his own findings, backing up Fleiss' 23- and 28-day theory. Though biorhythms have never been confirmed by scientific evidence, regular studies have been carried out that have yielded further cyclical theories.

**23 days—male • 28 days—female • 38 days—intuitional
43 days—aesthetic • 53 days—spiritual**

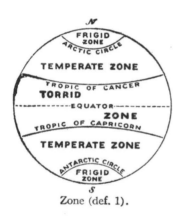

Zone (def. 1).

The axis of the planet Earth is 23.5 degrees to the vertical.

★

The tropics of Cancer and Capricorn are at 23.5° north and south respectively.

■ Michael Jordan, one of the all-time great basketball players, wore number 23 for the Chicago Bulls. The club retired his number after Jordan's first retirement in 1993, but he wasn't the only sportsman to have given the number 23 special significance.

Don Mattingly....................Baseball
(New York Yankees retired his shirt)

David Beckham .... Soccer
(while at Real Madrid in honor of his hero Michael Jordan)

Ryne Sandberg .... Baseball
(Chicago Cubs retired his shirt)

Shane Warne..................... Cricket
(Australia one-day side)

Marc-Vivien Foe.................. Soccer
(Manchester City retired his shirt after he collapsed and died on the pitch while playing for Cameroon in 2003)

89

THE NUMBER OF HOURS in a day—we have the ancient Egyptians and Babylonians (see **12**) to thank for choosing such a versatile number. For 24 is divisible by 1, 2, 3, 4, 6, 8 and 12. Which makes it very easy to divide the day into shorter periods.

ALPHA, BETA, GAMMA, DELTA, EPSILON, ZETA, ETA, THETA, IOTA, KAPPA, LAMBDA, MU, NU, XI, OMICRON, PI, RHO, SIGMA, TAU, UPSILON, PHI, CHI, PSI, OMEGA

A Alpha (al-fah)	B Beta (bay-tah)	Γ Gamma (gam-ah)	Δ Delta (del-ta)	E Epsilon (ep-si-lon)	Z Zeta (zay-tah)
H Eta (ay-tah)	Θ Theta (thay-tah)	I Iota (eye-o-tah)	K Kappa (cap-pah)	Λ Lambda (lamb-dah)	M Mu (mew)
N Nu (new)	Ξ Xi (zie)	O Omicron (om-e-cron)	Π Pi (pie)	P Rho (roe)	Σ Sigma (sig-mah)
T Tau (taw)	Y Upsilon (up-si-lon)	Φ Phi (fie)	X Chi (kie)	Ψ Psi (sigh)	Ω Omega (oh-may-gah)

The 24 letters of the Greek alphabet. This is the basis of all European alphabets, although different languages have adapted it in different ways, giving them different numbers of letters.

Italian .... 21 *(no J, K, W, X or Y)*
English 26
French... 26 *(as English)*
Spanish...................................... 29 *(adding CH-che, LL-elle, Ñ-eñe)*
German .... 30 *(adding Ä-ay, Ö-ooh, ß-ess-zett, Ü-uyuh)*
Dutch... 27 *(adding IJ-lange)*
Russian...33

✳

■ 24-carat gold is the purest form of gold. Other carat numbers denote the proportion of gold, for example 14 carat gold is 14 parts gold and 10 parts other metals.

Anything less than 10 carats cannot be sold as gold.

★

Homer's Odyssey *and* Iliad *are each divided into 24 books.*

The legend of King Arthur and the Knights of the Round Table takes many forms, none of which agree on the actual number of knights; it ranges from 12 up to several hundred. But in the Great Hall of Winchester Castle, in England, there is a huge depiction of the round table with the names of 25 knights including:

KING ARTHUR
SIR GALAHAD
SIR LANCELOT DU LAC
SIR GAWAIN
SIR PERCIVALE
SIR LIONEL
SIR TRISTRAM DE LYONES
SIR GARETH
SIR BEDIVERE
SIR BLEOBERIS
LA COTE MALE TAILE
SIR LUCAN

SIR PALOMIDES
SIR LAMORAK
SIR BORS DE GANIS
SIR SAFIR
SIR PELLEAS
SIR KAY
SIR ECTOR DE MARIS
SIR DAGONET
SIR TEGYR
SIR BRUNOR LE NOIR
LE BEL DESCONNEU
SIR ALYMERE
SIR MORDRED

★

The square of 5 and a quarter of 100, 25 is seen as a landmark number associated with silver. Couples who have been married for 25 years celebrate their Silver Wedding, and the 25-year reign of an English monarch is marked by a Silver Jubilee.

To stand for election to the United States House of Representatives you must be 25 years old.

25 is a popular card game in Eire.

There are 25 players on the roster in a Major League Baseball squad.

26 WEEKS is half a year. However, 26 x 2 x 7 = 364, leaving the year 1.25 days short. The first six months of the year only add up to 181 days (182 in a leap year), with 184 days in the second six.

26 cantons make up Switzerland.

★

26 is the numerical value of the four-letter name of God in Hebrew, Yod-He-Waw-He (see 27).

★

❏ At the age of 26 a man can no longer be drafted into the US army.

■ In a marathon, runners compete over 26 miles and 385 yards. This is supposed to be the distance run without stopping by the ancient Greek Pheidippides, who brought news of victory over the Persians in the Battle of Marathon to Athens. He then, reportedly, dropped dead on the spot.

27

THE 27 BOOKS OF THE NEW TESTAMENT

Matthew · Ephesians Hebrews · Mark Philippians · James · Luke Colossians · Peter I John · Thessalonians I Peter II · John I · Acts Thessalonians II Romans · Timothy I John II · Corinthians I Timothy II · John III Corinthians II · Titus Jude · Galatians Philemon · Revelation

*

27 seems to be the age of choice for the classic "live fast, die young" rock star. The "27 Club," as they've been dubbed, includes:

Brian Jones of the Rolling Stones (drowned)

Janis Joplin (drug overdose) **Jimi Hendrix** (drug overdose) **Kurt Cobain** (gunshot)

Jim Morrison (drug overdose)

Amy Winehouse (alcohol poisoning)

HIDDEN NUMBERS

THERE ARE 27 letters in the Hebrew alphabet, five of which (*kaf, mem, nun, fe* and *tzadi*) have a different pronunciation (*sofit*) when used at the end of a word. These letters also have a numerical value. The practice of matching letters to numbers in any language is called Gematria and it is particularly prevalent in Judaism, which believes that the holy book, the Torah, is the word of God given to Moses, and that God's messages are coded numerically in the Hebrew words of the Torah. Thus the number 7 (see **7**) is intrinsically linked with the Creation and is regarded as God's number.

Kabbalah is a mystical branch of Judaism which claims an authoritative interpretation of God's messages in the Torah, known only to the enlightened few. Traditional Kabbalists insist that it should only be studied within the strict observance of Jewish Law, otherwise it can lead to all sorts of misinterpretations which are tantamount to blasphemy. However, the widespread thirst for knowledge of the universe and what God wants us to know has led to a lucrative business in the teaching of Kabbalah, much of which may leave the student short of both enlightenment and cash.

28

THE MATCH of days and dates in the Gregorian calendar usually repeats in 28-year cycles, since there are seven days a week and a leap year every four years.

★

Another significant observation that the ancient astronomers made was that the planet Saturn returned to the same position in the night sky roughly every 28 years. Every 14 years its rings disappear, due to its angle in relation to earth, something which used to perplex even the most brilliant of stargazers. Saturn Return is an astrological term for a life change that is said to take place from the age of 28, due to the influence of Saturn. And the word Saturnine, meaning sluggish and morose, stems from the fact that, of the planets visible to the naked eye, Saturn takes the longest to orbit the Sun.

> **There are 28 dominoes in a standard "double six" set. A "double nine" contains 55 dominoes and a "double 12" contains 91.**

There are 28 letters in the Arabic alphabet.

✳

EXCLUDING WISDOMS, THE HUMAN MOUTH CONTAINS 28 TEETH.

✳

28 is a perfect number (see 6). Its divisors, 1+2+4+7+14, add up to 28.

My God, What Have We Done?

The number 29 will forever be associated with the atomic bombs dropped on Hiroshima and Nagasaki in World War II. The American plane that dropped the bombs was a B29.

Just before 8.15am on August 6 1945, B29 bomber *Enola Gay* dropped the bomb nicknamed "Little Boy" on Hiroshima. The force of the explosion was 15 kilotons, creating a fireball of more than 100,000°C that incinerated an estimated 66,000 people instantly. Three days later, at 11.02am, B29 bomber *Bocksca* dropped the "Fat Boy" bomb on Nagasaki, exploding with an even greater force of 22 kilotons.

Approximately 140,000 people died in the Hiroshima explosion, and around 70,000 in Nagasaki, nearly all of them civilians, and the numbers have increased dramatically since, possibly even doubled, due to the effects of the bombs, and not least cancers caused by the radiation. As he looked back on the destruction he had helped to cause, *Enola Gay*'s co-pilot Commander Robert Lewis was struck by a moment of remorse. He wrote in his log, "My God, what have we done?" Since Nagasaki, no country has used an atomic or nuclear weapon against another.

THERE ARE JUST OVER 29.5 DAYS BETWEEN V
FULL MOONS

★

Finland, Norway, Denmark and Turkey have 29 letters in their alphabet.

"My favorite poem is the one that starts, 'Thirty days hath September,' because it actually tells you something."

Groucho Marx

■ 30 is the product of the first three prime numbers: 1 x 2 x 3 x 5. Numbers that are the product of several consecutive primes, starting with 1, are called primordial. The first 5 primordials are:

2	1 x 2
6	1 x 2 x 3
30	1 x 2 x 3 x 5
210	1 x 2 x 3 x 5 x 7
2,310	1 x 2 x 3 x 5 x 7 x 11

❏ There is a rule of marketing that you should be able to "sell" any new idea or product in 30 seconds. This is known as an "elevator pitch," because you can get your idea across to a potential buyer in the time it takes to take the lift to their office.

■ 30mph is the normal urban speed limit in most parts of the world. In metric countries it is 50km/h, the equivalent of 31mph. In North America and Japan, 25mph is more common in congested areas.

Research has shown that a pedestrian hit by a car at 30mph has an 80 per cent chance of survival. At 40mph, this drops to just 10 per cent. At 20mph, the chance of survival is 95 per cent.

Club 18–30 is a holiday company, specializing in vacations for the young, free and single. The name reflects an unwritten belief that 30 marks the end of one's wild years.

■ The term "thirtysomething" became part of the English language in the 1980s, thanks to the hit TV series of that name. "Thirtysomethings" replaced the "yuppies" of the mid-1980s as the demographic phenomenon: baby boomers entering their thirties and having to confront adulthood.

In astrology, the Saturn Return ends at the age of 30 (see 28).

31 is a prime number, and so are these:

**331
3,331
33,331
333,331
3,333,331
33,333,331**

However, 333,333,331, and the eight further numbers in the sequence, are not prime, scotching a once-held theory that any number of 3s followed by a 1 would be a prime number.

$$333,333,331/17 = 19,607,843$$

★

31 IS A TYPE OF CARD GAME.

■ Seven months have 31 days: January, March, May, July, August, October, December.

✳ In chess there are 32 pieces on the board at the start of the game. There are also 32 squares of each color. Chess is a game riddled with numbers. These are all from competitive matches:

3—the fewest number of moves to have won a competitive match
72—most moves without a piece being captured
17—most consecutive captures
73—most consecutive moves by one piece
74—longest series of checks
141—highest total of checks in one game
15—highest number of pieces captured on the same square (a Black Hole)
40—the number of moves by each player in the first session.

■ **Jesus was crucified at the age of 32.**

❏ **Water freezes at 32° Fahrenheit/0°C.**

33 is a favorite number of conspiracy theorists. Dallas, where JFK was assass-inated, is located just south of the 33rd parallel. Baghdad is on the 33rd parallel.

MAKE MINE A 33

33 is the number of two popular beers. Heineken 33 is brewed in France, while Rolling Rock, brewed in the USA, features the number 33 on the label. This is said to stand for the number of words in the company's slogan: "Rolling Rock from the glass lined tanks of Old Latrobe, we tender this premium beer for your enjoyment as a tribute to your good taste. It comes from the mountain springs to you." The story goes that someone from the brewery wrote the number of words by the slogan when it was sent to the printer's, to give them an idea of the space required on the label. The labels came back with the 33 printed, and it stuck.

Including the bones of the coccyx, there are 33 vertebrae in the human spine.

" Then, the count added aloud, "Was his name ever known?" "Oh, yes; but only as No. 34." **"**

The Count of Monte Cristo
by Alexandre Dumas

During his time in prison, the book's hero, Edmond Danté, was known as number 34.

Miracle on 34th Street is a 1947 movie set in New York, featuring an 8-year-old Natalie Wood. It tells the story of a man who is declared insane because he claims to be the real Santa Claus, and ends up needing a miracle to convince the courts that he's for real. The film won three Oscars for Best Supporting Actor, Best Original Story and Best Screenplay.

ANTARCTICA
by numbers

Antarctica covers more of the globe than Europe but has a permanent population of zero. Only five per cent of Antarctica is actually made up of rock; the rest is ice.

Area
5,100,000sq.miles/13,209,000 sq.km

Percentage of global land mass
9%

Population (approx)
4,000 non-permanent scientists

Population density (per sq.km)
0.0000003

Highest point
Vinson Massif: 16,051ft/4,892m

Lowest point
8,327ft/2,538m bsl

Longest river
none

Highest recorded temperature
15°C /59°F (Vanda Station, Scott Coast, 1974)

Lowest recorded temperature
-89°C/-129°F (Vostok, 1983)

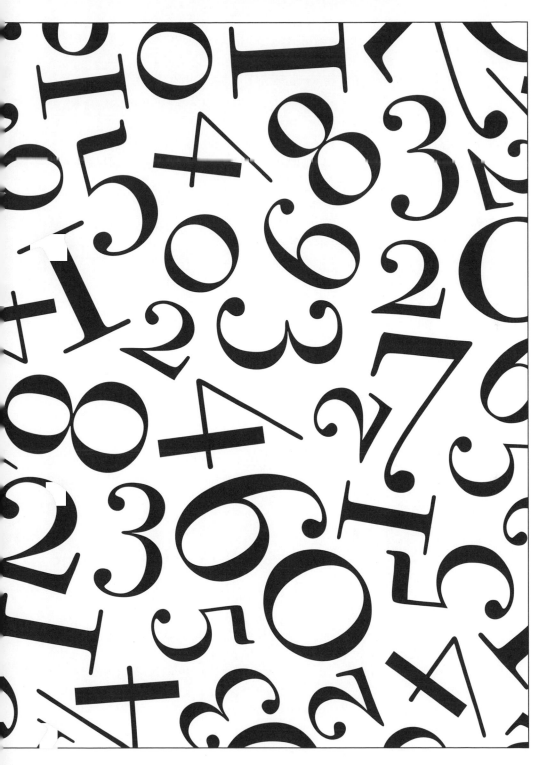

66Thirty-five is a very attractive age. London society is full of women of the very highest birth who have, of their own free choice, remained thirty-five for years.99

From *The Importance of Being Earnest* by Oscar Wilde

Anyone wishing to stand for President of the United States must be 35 years of age. To date, nobody has come within five years of this, the five youngest presidents being:

THEODORE ROOSEVELT 42
JOHN F. KENNEDY 43
BILL CLINTON 46
ULYSSES S. GRANT 46
GROVER CLEVELAND 47

35 years of marriage are celebrated with the coral wedding anniversary.

"Give me thirty-six hours and I'll give you a traitor"

This was the tag line of 1965 film *36 Hours*, starring James Garner. With D-day approaching and time running out for the Germans, they capture an American officer and attempt to convince him that the war is over so that he will divulge secrets.

✳

36 black keys on a piano

✳

JEWISH folklore tells of 36 "righteous people." According to the tale, at any given time there are 36 people on earth who hold this position but are unaware that they do so. The very thought that you might be one would be enough to prevent you from being one. As soon as one dies, another replaces him or her. Without the full complement of 36, the world would come undone.

37

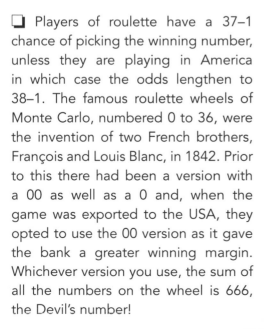

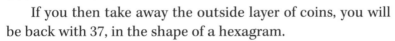

37°C (98.6°F) is body temperature in a healthy human being.

THOSE SEEKING spiritual significance in the number 37 will note that it is a combination of two highly mystical numbers, 3 and 7, that, along with 73, it figures in the Gematria (see **27**) of the first line of Genesis, and that these two numbers have an interesting geometrical relationship.

If you have 37 coins, you can lay them out in the shape of a hexagon. Numbers with this property are called "hex" numbers. If you then add coins to each side in the shape of six triangles to make a hexagram (six-pointed star), you will have a total of 73 coins.

If you then take away the outside layer of coins, you will be back with 37, in the shape of a hexagram.

☐ Players of roulette have a 37–1 chance of picking the winning number, unless they are playing in America in which case the odds lengthen to 38–1. The famous roulette wheels of Monte Carlo, numbered 0 to 36, were the invention of two French brothers, François and Louis Blanc, in 1842. Prior to this there had been a version with a 00 as well as a 0 and, when the game was exported to the USA, they opted to use the 00 version as it gave the bank a greater winning margin. Whichever version you use, the sum of all the numbers on the wheel is 666, the Devil's number!

"I'M 37! I'M NOT OLD!"
Peasant to King Arthur, *Monty Python and the Holy Grail*

37 is a psychologically random number (see **17**), and it therefore frequently crops up in fiction:

■ *Pulp Fiction* features a title card that says, "Nine minutes and thirty-seven seconds later."

■ In *Casablanca*, Rick is 37 years old.

38

DIRTY TRICKS

ACCORDING TO Arthur Schopenhauer, the 19th-century German philosopher, there are 38 ways to win an argument. Schopenhauer contended that the point of argument was not to be right, but to win, and drew up a list of 38 underhand tricks used to confound logic and reason. In doing so, he put together the toolbox for all politicians. Here are ten of the dirty tricks, which are very familiar today:

❏ Carry your opponent's proposition beyond its natural limits; exaggerate it.

❏ Use your opponent's beliefs against him.

❏ Confuse the issue by changing your opponent's words or what he or she seeks to prove.

❏ If your opponent presses you with a counter-proof, you will often be able to save yourself by advancing some subtle distinction.

❏ If your opponent has taken up a line of argument that will end in your defeat, you must not allow him to carry it to its conclusion.

❏ You may also puzzle and bewilder your opponent by mere bombast.

❏ When the audience consists of individuals (or a person) who are/is not an expert on a subject, you make an invalid objection to your opponent who seems to be defeated in the eyes of the audience.

❏ If you find that you are being beaten, you can create a diversion—that is, you can suddenly begin to talk of something else, as though it had a bearing on the matter in dispute.

❏ Instead of working on an opponent's intellect or the rigor of his arguments, work on his motive.

❏ Become personal, insulting and rude as soon as you perceive that your opponent has the upper hand.

❝"Have you ever heard of the 39 Steps?"
"No. What's that, a pub?"❞

In fact, the 39 Steps is the code in John Buchan's gripping spy novel, made into a film by Alfred Hitchcock in 1935 (from which this dialogue is taken), and remade in 1959 and 1978. Buchan took the idea of 39 steps from a feature on a cliff path in Kent, near to where he was staying when he wrote the book.

■ Before it was rebuilt, Wembley Stadium in London, the "Mecca" of soccer, also had 39 steps. They led from pitch level to the Royal Box, and teams had to climb them to receive their medals after cup finals. The new stadium has 107 steps to the Royal Box, so, given the physical state of most players by the time they've played a game at Wembley, some sort of mechanical stairlift may be in order.

39 is the sum of five consecutive prime numbers: 3 + 5 + 7 + 11 + 13. It is also the product of the first and last of these numbers (3 x 13), which actually makes it a rarity. 10 is the first and the next is 155. Despite this, it was singled out by David Wells in his book *The Penguin Dictionary of Curious and Interesting Numbers* as the "smallest uninteresting number," by which he meant that there was nothing significant about it at all. It has been argued that being the first uninteresting number is of interest in itself, and therefore there can be no such thing as the first uninteresting number.

★

"Age is strictly a case of mind over matter. If you don"t mind, it doesn't matter."

Jack Benny'comedian.

Benny always kept his age at 39, because "There's nothing funny about 40." By the time he died he had celebrated his 39th birthday 41 times.

> # The Anglican Church is founded on the 39 Articles of Religion.

40 crops up with alarming regularity in the Christian and Muslim scriptures and now marks the period of Lent, but it may not have been meant as such a specific number. A likely explanation is that 40 is used in Middle Eastern culture to mean "a lot." For example, *Ali Baba and the Forty Thieves.*

LIFE BEGINS AT 40

THIS WOULD have rung hollow in Roman times, when the average life expectancy at birth was a mere 22. In fact, even in the USA as recently as the turn of the 19th/20th century, life expectancy was just 49. However, these low figures were largely affected by the high rate of infant mortality (and regular wars can't have helped). If you lived to see 40, statistically you could look forward to another 20 years on average. Hardly party time, but then the phrase has always been open to question. Philosopher **Carl Jung** called 40 the "noon" of life, trigger for mid-life crisis. For those who planned families young, it could signify getting shot of the kids and getting your life back.

"If you're not a liberal at twenty you have no heart; if you're not a conservative at forty, you have no brain."

Winston Churchill

"She said she was approaching forty, and I couldn't help wondering from what direction."

BOB HOPE

"*Forty is the old age of youth; fifty is the youth of old age.*"

Victor Hugo

WD-40— "*the can with a thousand uses.*" *Invented by Norm Larsen, the WD stands for water displacement and the 40 for the fact that it was Larsen's 40th attempt to get the formula right.*

40 acres and a mule

After the American Civil War, General Sherman ordered the distribution of land to freed black slaves in lots of 40 acres. The process was summed up colloquially as "40 acres and a mule" which was used by black American film director Spike Lee as the name of his production company.

"*A man has more character in his face at forty than at twenty—he has suffered longer.*"

Mae West

41

THIS PRIME NUMBER, forms an unlikely connection between Charlton Heston and Mozart. It was Charlton Heston's number as a galley slave in the movie *Ben Hur*, for which he won an Oscar, and it's the number of Mozart's last symphony, later dubbed the Jupiter Symphony. Mozart died at just 35, but in that time he wrote over 600 musical works. In addition to numerous duets, quartets, quintets, serenades, concertos and religious pieces, he wrote:

41 SOLO PIANO PIECES
41 SYMPHONIES
36 VIOLIN SONATAS
27 PIANO CONCERTOS
23 OPERAS

Heston has made over 80 movies to date and was the President of the Screen Actors Guild from 1965 to 1971.

THE FIRST BOOK ever printed, Joseph Gutenberg's Bible (1454), contains 42 lines to the page. Only 16 complete copies of the *Gutenberg Bible* remain, from a total print run of 180. Interestingly, Gutenberg altered his type layout during the production, having started off with 40 lines to the page. The print run was increased after the first versions had been printed, and the reprints were all reset with 42 lines per page.

ELVIS PRESLEY DIED AT THE AGE OF 42.

"Forty-two! Is that all you've got to show for seven and a half million years' work?"

■ As every fan of Douglas Adams' sci-fi satire *The Hitchhiker's Guide to the Galaxy* knows, 42 is the "ultimate answer to life, the universe and everything." It takes a massively powerful computer, Deep Thought, 7.5 million years to work it out, and even then it can't provide us with the ultimate question. Disciples of the book have formulated countless theories on the meaning of 42 and why Adams chose it. The fact is he picked it at random.

★

42 was a significant number for the Egyptians. They believed that the dead would have their heart weighed by the god Osiris and would stand trial before 42 judges, where they would have to deny 42 sins.

The Beatles' last performance lasted 42 minutes. It took place unannounced on the roof of the Apple Records building in London on January 30 1969, halting traffic as passers-by stopped to try and catch the impromptu performance.

SURPRISING to find two prime numbers so close together this far up the numerical scale, but 43 swiftly follows 41. Two prime numbers like this that are two apart are called Twin Primes, and they do, in fact, appear in the billions ... and possibly beyond. Whether there are an infinite number of Twin Primes has yet to be proven. Here are the first 10. Notice that 5 appears twice.

| 3, 5 | 11, 13 | 29, 31 | 59, 61 | 101, 103 |
| 5, 7 | 17, 19 | 41, 43 | 71, 73 | 107, 109 |

★

Hancock's 43 Minutes *was a 1957 Christmas Special of* Hancock's Half Hour, *featuring the great British comedian Tony Hancock.*

★

■ **43AD was the year of the Roman conquest of Britain. Not to be mistaken for Julius Caesar's invasions of 55 and 54BC, when he only gained control of a small part of the country.**

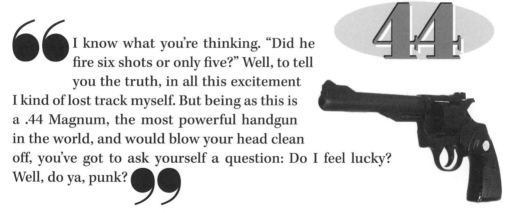

"I know what you're thinking. "Did he fire six shots or only five?" Well, to tell you the truth, in all this excitement I kind of lost track myself. But being as this is a .44 Magnum, the most powerful handgun in the world, and would blow your head clean off, you've got to ask yourself a question: Do I feel lucky? Well, do ya, punk?"

❑ "Dirty" Harry Callahan, played by Clint Eastwood, used a .44 Magnum, a large-bore cartridge designed for revolvers and rifles. It was also used by Travis Bickle (Robert de Niro) in *Taxi Driver*. But despite its immortalization in these two 1970s cinema classics, it is not "the most powerful handgun in the world." This is the Smith & Wesson Model 500.

A little like Douglas Adams and 42, 44 was the number chosen by Mark Twain to represent the character of Satan in his unfinished novel, The Mysterious Stranger.

"Brimful of Asha on the 45
Well, it's a brimful of Asha on the 45"

"Brimful of Asha" by Cornershop

In the days of vinyl records, singles were
called 45s because they spun at 45rpm (see **78**).

■ The Colt 45 is one of the most famous revolvers ever
made, a classic of the Wild West. It was designed for
the US Cavalry in the late 19th century and assumed
the ironic nickname of the "Peacemaker."

Boxer George Foreman was 45 when he became the oldest man ever
to win the World Heavyweight title. "Big George" is now a born-again
Christian and has named each one of his five sons George, "so they
know who their father is," an impulse which has also led to one of his
daughters being called Georgetta. Since 1995, he has sold over 100
million of his fat-reducing grills.

46 was John F Kennedy's age when he
was assassinated in 1963. It was also the age
of golfer Jack Nicklaus when he became
the oldest-ever winner of the US
Masters in 1986.

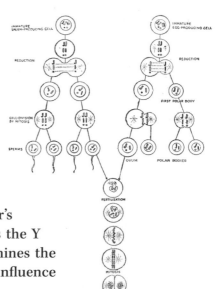

46 is the number of chromosomes in
a human cell. These are arranged
into 23 pairs and children inherit one
from each pair from each parent. One of
the 23 pairs is the sex chomosomes, which
determine gender. Females have two X sex
chromosomes and males have an X and a
Y. If the child inherits the X from the father's
pair, it will be female (XX), and if it inherits the Y
it will be male (XY). Thus, the father determines the
sex of the child. And that's about where his influence
ends, some say.

"Well, number 47 said to number 3
You're the cutest jailbird I ever did see."

"Jailhouse Rock" by Elvis Presley

The AK-47 is the most widely used assault rifle in the world. The 47 refers to the year of its commission, 1947.

A concert harp has 47 strings, three octaves below middle C and three-and-a-half above.

★

THE 47 SOCIETY, formed at Pomona College, California, have, for their own amusement, spread the notion that 47 occurs with unnaturally high frequency. The idea is said to have originated in 1964 with a Professor Donald Bentley, who somehow proved that all numbers equal 47. Several Pomona graduates used their influence to infiltrate 47s into popular culture, none more so than Joe Menosky. As a writer on *Star Trek: The Next Generation*, *Voyager* and *Deep Space Nine*, he, so the story goes, set out to work the number 47 into every episode he could. He also spread the habit to other writers, with the result that 47 (or in some cases 74) is alleged to figure in just about every episode. Trekkies have listed incidences of 47 on various websites. *47* is also the title of the unfinished fictional film in David Lynch's *Inland Empire*.

★

48

"The man who is a pessimist before forty-eight knows too much; if he is an optimist after it he knows too little."

Mark Twain

"48 hours needs 48 thrills."

"48 Thrills" by The Clash

✻

48 hours is commonly used as a more dramatic way of saying two days, particularly the weekend. It sets the clock ticking.

❏ *48 Hours* is a 1982 film starring Eddie Murphy and Nick Nolte as a convicted criminal and a cop respectively, forced to work together to track down a killer. And they've only got two days.

✻

48 is the smallest number with exactly ten divisors: 1, 2, 3, 4, 6, 8, 12, 16, 24 and 48

✻

■ The Forty & Eight is a society of American war veterans, formed in 1920 after the First World War. It takes its name from the rail wagons used to carry troops to the front in France. These were marked with the figures 40/8, signifying that they had the capacity to carry 40 men or eight horses.

❏ *The Forty-Eight* is another name for Johann Sebastian Bach's *Well-Tempered Clavier*. It consists of 48 pieces: a prelude and a fugue in each of the major and minor keys.

49

49 days are significant in Buddhism. It is the time Buddha spent meditating under the sacred tree when he received enlightenment. The soul is said to wander for 49 days between death and rebirth. And the Buddhist funeral rites last for 49 days.

■ When heavyweight boxing champion Rocky Marciano retired in 1956, his record was won 49, lost 0. It remains the longest undefeated streak in heavyweight boxing and Marciano remains the only World Heavyweight Champion to go undefeated throughout his career. Forty-three of his wins were knock-outs.

The '49ers

THE CALIFORNIA GOLD RUSH began in 1848 on a farm called Sutter's Mill, in the foothills of the Sierra Nevada. Word of the find spread slowly at first, as those in the know tried to keep the secret to themselves, but by 1849 the stampede had begun in earnest.

News of gold in California had spread around the world, and prospectors came from all points of the compass to try their luck. These were the '49ers. As many as 90,000 gold diggers arrived in California that year, causing a radical change in the make-up of California and particularly the towns near that first strike. San Francisco had been a small seaport up until then. Almost overnight its population boomed from around 1,000 to nearer 25,000 as ships sailed in from all over the world.

It's been estimated that around $7billion worth of gold was recovered in the first five years of the Gold Rush, making a lot of people very rich. However, it was not good news for everybody. One in twelve of the '49ers died in the process of seeking their fortune. The local Native American population, their lands invaded, their people attacked and starved, dwindled from around 150,000 to about 30,000.

But the Gold Rush was the making of California as we know it today, the epitome of the "new American dream," a symbol of get-rich-quick culture. In 1850 it was officially recognized as the 31st state of the Union and today it is known as the Golden State. And of course, San Francisco's NFL team is called the '49ers.

THE "NIFTY FIFTIES" gave us rock'n'roll, James Dean, Elvis, Marilyn Monroe, the Cold War, the Suez Crisis, the Korean War, Mods and Rockers, the jive, the beehive, the diner and the teenager. The 1950s marked a watershed in 20th-century culture, either because it was the first decade since World War II or because it was the halfway point in the century. Interestingly, the 1850s also saw major challenges to the status quo. Charles Darwin published *The Origin of Species*, Neanderthal Man was discovered, war raged in the Crimea and the Indian Mutiny signalled a major threat to the British Empire.

"The problem is all inside your head, she said to me
The answer is easy if you take it logically
I'd like to help you in your struggle to be free
There must be 50 ways to leave your lover"

"50 Ways to Leave Your Lover" by Paul Simon

■ THE HALF-CENTURY is seen as a significant landmark in age, sporting achievement, marriage (the golden wedding anniversary) and monarchy (golden jubilee). The term "jubilee" originally comes from a Hebrew word for a ram's horn. The horn was blown to signal a celebration every 50 years, when slaves were given their freedom.

FOR 50 is the fourth Roman numeral. Of all the Roman numerals, only C (*centum*) for 100 and M (*mille*) for 1,000 actually stand for the Roman word. All the others, I, V, X, L, and D, evolved from the shapes of notches cut in sticks as part of a primitive counting system.

Hawaii is the 50th state of the USA, which explains the name of the 1970s TV show Hawaii Five-0.

❏ **50 Gates of Wisdom** is a concept in Jewish Kabbalistic belief, stemming from the Bible story of the Israelites' exodus from Egypt. Before Moses received the Torah on Mount Sinai, they wandered for 49 days, during which time they were forced into a lot of introspection. Today, believers regard the 49 days, plus the prophetic 50th day, as a time of self-evaluation.

"At 50, everyone has the face he deserves."

George Orwell

"Sex appeal is fifty per cent what you've got and fifty per cent what people think you've got."

Sophia Loren

✻

■ **THE 50-YEAR RULE** states that an item falls out of copyright 50 years after its creation. It applies to music recordings, although there is a campaign to extend this to 95 years. In many countries, copyright applies in many fields for 50 years beyond the death of the creator of the work. This rule varies for different places and different media, so check!

50 eggs

The film *Cool Hand Luke*, starring Paul Newman as Lucas Jackson, features a scene in which Jackson bets he can eat 50 hardboiled eggs in one hour. Up until 2003, the world record for eating hardboiled eggs stood at just 38 in eight minutes. But then Sonya Thomas, a petite American of Korean descent, put away 65 in just 6 minutes 40 seconds. Not only did she whip Cool Hand Luke for speed, but for sheer quantity it takes some beating.

Eating contests have become a global "sport," all under the jurisdiction of the International Federation of Competitive Eating. Here are some of the records to date:

Baked beans	6lb in 1 min 48 secs
Butter	7 x ¼lb sticks in 5 mins
Cabbage	6lb 9oz in 9 mins
Chicken nuggets	80 in 5 mins
Cow brains	57 in 15 mins (17.7lb total)
Glazed doughnuts	49 in 8 mins
Vanilla ice cream	1 gallon 9 oz in 12 mins
Mayonnaise	4 x 32oz bowls in 8 mins
Onions	8.5oz in 1 min
Oysters	552 in 10 mins

51

> ❝A DEMOCRACY IS NOTHING MORE THAN MOB RULE,
> WHERE FIFTY-ONE PER CENT OF THE PEOPLE MAY
> TAKE AWAY THE RIGHTS OF THE OTHER FORTY-NINE.❞
>
> Thomas Jefferson

THE PARKER 51

THE PARKER 51 is one of the most famous pens ever made. It was designed to use quick-dry ink, a new invention by Parker that required a new type of plastic to be used for the body of the pen. The Parker 51 was made with Lucite, which, unlike other plastics, did not corrode on contact with the ink. The ink and the pen were branded 51 because they were developed for use in the company's 51st year (1939). It was launched in 1941, and huge demand for the pen built up during the war, because of manufacturing restrictions, and the numerical name made it easy to market worldwide. When the restrictions were lifted and Parker was able to meet the demand, the 51 sold in its millions, right up until 1972, netting the company in excess of £400million.

★

51 is a brand of pastis, the French anise liqueur made by Pernod Ricard. Its name is taken from the year of its launch, 1951.

★

52

THE 52-CARD DECK

THE PASSION for playing cards as we know them today caught hold in Europe in the late 14th century. The concept originated in China, but then the 52-card deck, containing four suits of 13 with three "picture" cards per suit, was introduced to Europe from Egypt around that time. There followed a period of experimentation, with different suit names and emblems, extra suits, cards added to each suit, or the value of the "picture" cards swapped around. It was the French who settled on the now familiar suits of hearts, diamonds, spades and clubs. They called them *coeurs*, *carreaux*, *piques* and *trèfles*, the English "spades" coming from the Italian "spade" meaning sword, a sword having been a former emblem of the suit. The French also instigated the notion of "aces high," whereby what had always been the lowest-value card now became the highest.

Jokers were introduced in America in the 19th century and spread back to Europe along with poker. The practice of printing the manufacturer's trademark on the ace of spades dates back to a British law passed by James I, stipulating that tax should be paid on the manufacture and sale of playing cards, and proof of payment stamped on the ace of spades.

The 52-card deck is known as a poker or Anglo-American deck. Elsewhere in the world the format varies, with different suits still used, and even different numbers of cards in each suit. In Russia, for example, numbers 1 to 5 are discarded.

❝Love is...
Two minutes and 52 seconds of squelching noises.❞

According to John Lydon (aka Johnny Rotten)

"As I see it, there is not much difference between being sixty-three and fifty-three: whereas when I was fifty-three I felt at a staggering distance from forty-three."

Simone de Beauvoir, French feminist philosopher and writer

❏ 53 was the number of **Herbie**, the racing Volkswagen Beetle with a mind of its own, which has starred in five movies:

THE LOVE BUG (1968)
HERBIE RIDES AGAIN (1974)
HERBIE GOES TO MONTE CARLO (1977)
HERBIE GOES BANANAS (1980)
HERBIE: FULLY LOADED (2005)

THERE ARE 54 colored squares on a standard Rubik's Cube (six faces of nine squares). Other Rubik's Cubes have four, 16 and 25 squares per face, and independent makers have even gone up to 121, but the original, as invented by Hungarian sculptor

Erno Rubik in 1974, had nine. The Rubik's Cube was launched worldwide in 1980 and sold over 100,000,000 in its first two years. It is thought to be the best-selling "toy" ever, with total sales somewhere in excess of 250 million. Despite there being 43,252,003,274,489,856,000 possible positions for a nine by nine cube, there is a method that will always solve it in 27 moves or less.

■ **Studio 54** was the venue at the heart of the New York disco explosion in the late 1970s. Located at 254 West 54th Street in Manhattan, it ran for nine years from April 1977 to March 1986, pulling in a glamorous mixture of celebrities and beautiful young things. In 1998 a poorly received film was made about the club.

AFRICA IS MADE UP OF 55 COUNTRIES:

ALGERIA	CÔTE D'IVOIRE	LIBYA	SENEGAL
ANGOLA	DJIBOUTI	MADAGASCAR	SEYCHELLES
BENIN	EGYPT	MALAWI	SIERRA LEONE
BOTSWANA	EQUATORIAL	MALI	SOMALIA
BURKINA FASO	GUINEA	MAURITANIA	SOUTH AFRICA
BURUNDI	ERITREA	MAURITIUS	SUDAN
CAMEROON	ETHIOPIA	MOROCCO	SWAZILAND
CAPE VERDE	GABON	MOZAMBIQUE	TANZANIA
CENTRAL	GAMBIA	NAMIBIA	TOGO
AFRICAN REP.	GHANA	NIGER	TUNISIA
CHAD	GUINEA	NIGERIA	UGANDA
COMOROS	GUINEA-BISSAU	REUNION	WESTERN SAHARA
CONGO	KENYA	RWANDA	ZAMBIA
(BRAZZAVILLE)	LESOTHO	SAO TOME &	ZIMBABWE
CONGO (DRC)	LIBERIA	PRINCIPE	

55 crores was the sum of rupees that caused an international incident between India and Pakistan in 1947. A crore is part of a Hindu numbering system in which a lakh is 100,000 and a crore is 100 lakhs, equivalent to 10 million. When an amount is written, the commas are positioned differently to the norm, demarking lakhs and crores, rather than millions, billions etc. So 55 crores would be written 55,00,00,000.

During the India-Pakistan partition, an agreement had been made for India to pay Rs 75 crores to Pakistan in settlement of assets, of which 20 was paid without any fuss. But when insurgents, believed to be supported by the Pakistani government, invaded part of Kashmir, the Indian government decided to hold back the remaining 55 crores. Gandhi strongly urged the Indian government to pay the money, and this stance is seen in some quarters as a contributory factor towards his assassination on January 30 1948.

AFRICA
by numbers

Africa covers an area that is about 20 per cent of the world's total land mass, but only about 10 per cent of the world population live there. It boasts the hottest temperatures and the longest river.

Area
11,608,000sq.miles/
30,065,000sq.km

Percentage of global land mass
20%

Population (approx)
900,000,000

Population density (per sq.km)
30

Highest point
Mt Kilimanjaro: 19,340ft/5,895m

Lowest point
Lake Assal, Djibouti: 512ft/156m bsl

Longest river
The Nile: 4,157 miles/6,690km

Highest recorded temperature
58°C/136°F (El Azizia, Libya, 1922)

Lowest recorded temperature
-24°C/-11°F (Ifrane, Morocco, 1935)

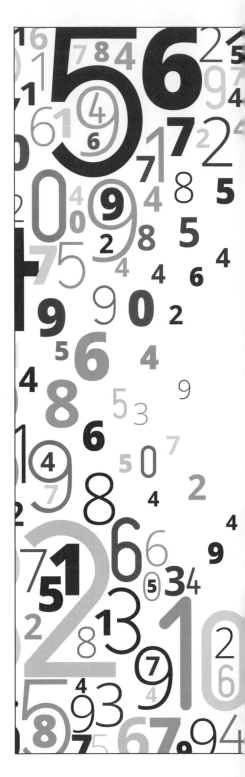

THE UNITED STATES Declaration of Independence in 1776 was signed by 56 men representing 13 states. Most famous among them were Benjamin Franklin of Pennsylvania, John Hancock of Massachusetts, and two future Presidents, John Adams of Massachusetts and Thomas Jefferson of Virginia. The 13 states were:

DELAWARE	NEW YORK	NEW JERSEY
PENNSYLVANIA	GEORGIA	CONNECTICUT
MASSACHUSETTS	VIRGINIA	MARYLAND
NEW HAMPSHIRE	NORTH CAROLINA	
RHODE ISLAND	SOUTH CAROLINA	

✳ **56lbs are half a hundredweight.** ✳

56.5 inches is the standard railway gauge throughout the world. Better expressed as 4ft 8½in, this imperial measurement is a legacy of the Romans, via the British Empire.

When the Romans built their chariots, they needed to set the wheels a standard width apart, allowing two horses side by side to be yoked between them. One knock-on effect of wheeled transport was the ruts it caused in the roads, which could get fairly deep. Any vehicle with wheel that didn't fit in these ruts ran the risk of being pulled apart. So a standard was set which then spread throughout the Roman Empire as they built more and more roads.

Fast-forward to industrial Britain in 1814, when George Stephenson invented his first locomotive. Stephenson was familiar with the horse-drawn cart tracks that serviced the coalmines around his native Newcastle and designed his engines to operate on this

track. The gauge, which is given his name, was 4ft 8½in.

As the railways spread, it became obvious that a standard needed to be set so that they could all converge, and in 1845 the British Government settled on 4ft 8½in. The one objection had come from the brilliant engineer Isambard Kingdom Brunel, who claimed that 4ft 0¼in was the optimum gauge for railways, and he was probably right, but he was outnumbered.

As railways grew up around the world, it was often British engineers who built them and British-built locomotives running on them, so the standard spread. Today, around 60 per cent of the world's railways run on 4ft 8½in gauge track.

IT'S A TESTAMENT to the power of marketing that the number 57 is most readily associated with the varieties of Heinz sauces. So what are the 57 varieties? Allegedly there's no such thing. Heinz picked the number at random and it stuck. With a vengeance.

You can reach Heinz corporate headquarters in Pittsburgh by dialling +1 412 237 5757, or by writing to PO Box 57, Pittsburgh, PA.

58 IS THE NUMBER of letters in the name of the famous Welsh town of Llanfairpwllgwyngyllgogerychwyrndrobwllllantysiliogogogoch. The name is said to have been made up as a joke to confuse the English, but it does have a meaning: "The church of Mary in the hollow of white hazel trees near the rapid whirlpool by St Tysilio's of the red cave." Indeed, it has its own website, which boasts "the longest domain name in the world." The site then qualifies this by pointing out that there is a longer domain name, for a site serving the upper part of the village, and it looks like this: www.Llanfairpwllgwyngyllgogerychwyrndrobwllllantysiliogogogochuchaf.com

While Llanfair (or LlanfairPG, as it is known for ease) is the longest place name in the UK, the *Guinness Book of Records* cites the New Zealand town of Tetaumatawhakatangihangakoauaotamateaurehaeaturipukapihimaungahoronuku pokaiwhenuaakitanarahu as the world's longest official place name with 92 letters. The name means the "place where Tamatea, the man with the big knees, who slid, climbed and swallowed mountains, known as land-eater, played his flute to his loved one."

The longest unofficial name, with 167 characters, is in Thailand: Krungthep Mahanakhon Bovorn Ratanakosin Mahintharayutthaya Mahadilokpop Noparatratchathani Burirom Udomratchanivet Mahasathan Amornpiman Avatarnsathit Sakkathattiyavisnukarmprasit, also known as Bangkok. This, rather implausibly, translates as "city of angels." Like LA.

✳

In 1922, a temperature of 58°C (136.4°F) was recorded in Libya. This still stands as the highest temperature ever recorded on earth.

59

" *Life I love you, all is groovy* **"**

"The 59th Street Bridge Song"
("Feelin' Groovy") by Simon & Garfunkel

The bridge, also known as The Queensboro Bridge, spans the East River in New York, connecting Manhattan with Long Island.

■ *For every 59 rotations of the Earth, Mercury rotates once.*

60

IF YOU REMEMBER the Sixties, as the saying goes, you probably weren't there. The most overhyped decade in the history of mankind, the Swinging Sixties was undeniably a period of great change. It was an era in which sex, drugs and rock'n'roll mixed with revolution, war and assassination. And at the end of it all, they put a man on the moon. For those caught up in the protests or the pop charts, it was the greatest decade ever. For everyone else, it was just another ten years.

60mph is the national speed limit in the UK, where speed limits were invented. The first speed limit for automobiles was 10mph in 1861, but four years later this was reduced to 4mph when it was seen what damage could be done by a two-ton car hitting a horse. 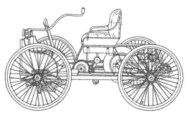 For further safety, a man had to walk 60 yards in front of the car carrying a red flag. This law prevailed for 31 years, before the speed limit was increased to a hair-raising 14mph in 1896.

SUMERIAN 60S

60 was a very important number for the early civilizations. The Sumerians, who prevailed in the Middle East around 7,000 years ago, had a base ten counting system which they used to create measurements that we still use today.

60 is a very versatile number for mathematicians. It can be divided by all the numbers from 1 to 6, as well as by 10, 12, 15, 20, 30 and, of course, 60. The Sumerians used this versatility to divide up the sky into six "houses" each subdivided into units of 60. This was depicted as a circle divided into 360°, with each degree then subdivided into 60 minutes (small parts) and each minute into 60 seconds (secondary small parts). This is the system used in geometry today, as well, of course, as in the measurement of time.

The Sumerians would have recognized that 60° is the internal angle of an equilateral triangle.

The importance of 60 in counting is still evident today. In French, the multiples of 10 beyond 60 do not have a name of their own. Instead, numbers between 60 and eighty are counted as additions to 60, e.g. 75 is *soixante-quinze* (sixty-fifteen). Similarly, in old English, but still used occasionally today, 60 is "three-score," 70 "three score and ten."

The Chinese lunar calendar has a 60-year cycle called the "stem-branch" cycle.

❋

In old English money, 60 pence was called a crown.

❋

There are 60 pegs or marbles in Chinese checkers.

❋

60 years of marriage are celebrated with a diamond anniversary.

ON JULY 9 2005, **Danny Way** performed the remarkable feat of becoming the first person in the world to jump the Great Wall of China on a skateboard. Unaided by any sort of motor, Way had to gather speed by rolling down a huge ramp until he was traveling at nearly 50mph. On clearing the wall, the distance of the leap was measured at 61 feet.

✳

> **"**Now the fifth daughter on the twelfth night
> Told the first father that things weren't right
> My complexion she said is much too white
> He said come here and step into the light
>   he says hmm you're right
> Let me tell the second mother this has been done
> But the second mother was with the seventh son
> And they were both out on Highway 61**"**

"Highway 61 Revisited" by Bob Dylan

The road in question, known as "The Blues Highway," runs from New Orleans, up the Mississippi and on to the Canadian border. Elvis Presley grew up in a housing project along Highway 61, and it was in a motel just off the road that Martin Luther King Jr was killed by a sniper's bullet.

**Wallace and Gromit live at
62 West Wallaby Street.**

★

■ *July and August have a total of 62 days, the
highest number of days in consecutive months.*

63 years is the record reign of any queen in history. Between 1837 and 1901, Queen Victoria reigned for 63 years, seven months and two days.

64 IS ANOTHER versatile number. It is the smallest number with exactly seven divisors: 1, 2, 4, 8, 16, 32 and 64. Other than 1, it is the first number to be both a square and a cube. It is also a sixth power.

$$2^6 = 64$$
$$4^3 = 64$$
$$8^2 = 64$$

❝*When I grow older, losing my hair*
Many years from now...❞

"When I'm 64" by the Beatles

■ "The 64 dollar question" means "the key question." It comes from a gameshow called *Take It Or Leave It* which ran on US radio in the 1940s. $64 dollars was the jackpot. Today, you're more likely to hear, "That's the 64 million dollar question," because $64 is too small a sum to get excited about.

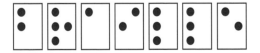

❏ **Braille,** the reading and writing system for the blind, which was devised by Louis Braille in 1821, consisted of 64 characters. Using a system of six dots arranged in two columns of three, each character was made up of a combination of different numbers of these dots raised from the paper so that they could be felt by the fingers. Braille now uses an 8-dot system, which gives it greater scope.

■ The 64 Arts are the different positions detailed in the *Kama Sutra,* the original lover's guide. Believed to have been written between the first and sixth centuries AD by a religious student called Vatsyayana, the *Kama Sutra* is a social study of Indian culture and sexuality. And while Vatsyayana writes that there are eight ways of making love, each with eight possible positions, the proportion of the *Kama Sutra* that details the various positions is only a small part within 36 chapters.

64 hexagrams in the I Ching
(see 8)

✳

64 squares on a chess board

❝Retirement at sixty-five is ridiculous.
When I was sixty-five, I still had pimples.**❞**

George Burns, comedian

65 is no longer the statutory retirement age for men in the USA, having been raised to 67 for people born later than 1959. 65 remains the men's retirement age in many European countries and elsewhere, but the rise in life expectancy looks likely to change all that.

The traditional materials associated with wedding anniversaries mysteriously run out at 65. While it would have been deemed unlikely for any marriage to last 65 years, the association does resume at 70. The traditional anniversary materials are:

1	Paper	8	Bronze	14	Ivory	45	Sapphire
2	Cotton	9	Pottery	15	Crystal	50	Gold
3	Leather	10	Tin, Aluminum	20	China	55	Emerald
4	Linen			25	Silver	60	Diamond
5	Wood	11	Steel	30	Pearl	70	Platinum
6	Iron	12	Silk	35	Coral	75	Diamonds again
7	Wool	13	Lace	40	Ruby		

✻
But 65 isn't all bad news. If you like magic squares, 65 is the constant of a 5 by 5.
✻

17	24	1	8	15
23	5	7	14	16
4	6	13	20	22
10	12	19	21	3
11	18	25	2	9

WITH apologies to Bob Dylan (see **61**), 66 is the number of the most famous road in rock. While Dylan's Highway 61 may have featured in more songs, Route 66 is the most recognized and the most covered, ever since Nat King Cole had a hit with it in 1946. The road itself runs from Chicago to LA—"more than two thousand miles all the way"—cutting through six states and, as the song explains, taking in a number of places that otherwise would probably have remained anonymous.

66 is seen as a significant number in English history, as it appears in three key dates:

1066—*the Norman Conquest*

1666—*the Great Fire of London*

1966—*England won the World Cup*

ANOTHER PRIME NUMBER with a 7 in it, 67 forms an interesting sequence of multiplications:

$$67 \times 67 = 4,489$$
$$667 \times 667 = 444,889$$
$$6,667 \times 6,667 = 44,448,889$$

etc., etc.

■ **68 years** passed between the formation of the Third French Republic in 1871 and the German invasion in 1939.

Sixty-eight Publishers is a Canadian publishing company founded in 1971 by Czech exiles. Sixty-eight refers to 1968, the year of the Soviet invasion of Czechoslovakia. It was a year of violent civil unrest around the world: student uprisings in Paris and Mexico City; civil rights upheavals in the USA; anti-Vietnam War demonstrations in London and America; the assassinations of Martin Luther King Jr and Robert F. Kennedy.

NORTH AMERICA
by numbers

North America comprises just two countries but, at 93,000kWh per capita per year, its people consume more than twice as much energy per head as any other continent, and four times the global average.

Area
9,365,000sq.miles/24,256,000sq.km

Percentage of global land mass
16%

Population (approx)
515,000,000

Population density (per sq.km)
21

Highest point
Mount McKinley: 20,320ft/6,194m

Lowest point
Death Valley, CA: 282ft/86m bsl

Longest river
Mississippi/Missouri 3,877 miles/6,236km

Highest recorded temperature
57°C/134°F (Death Valley, CA, 1913)

Lowest recorded temperature
-63°C/-81.4°F (Snag, Canada, 1947)

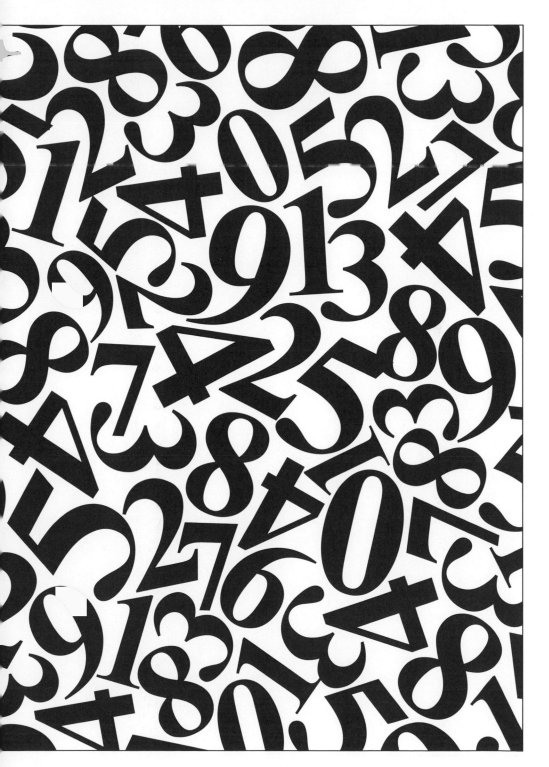

69 IS THE POPULAR NAME for a particular type of coupling. The *Kama Sutra* calls it the congress of the crow, but as a visual device, "69" is a bit more explicit about what's involved. The term (*soixante-neuf*) was coined in France in the 1800s.

George Washington received 69 votes to become the first President of the United States.

■ **69** is the record number of children born to one mother, 67 of whom survived. The record was set by a woman from Shuya in Russia in the 18th century: 27 pregnancies of which none were singletons.

- ❏ 16 pairs of twins   (32)
- ❏ 7 sets of triplets   (21)
- ❏ 4 sets of quadruplets   (16)

Vat 69 is a brand of whisky.

" *The days of our years are threescore years and ten;*

and if by reason of strength they be fourscore years,

yet is their strength labor and sorrow;

for it is soon cut off, and we fly away **"**

Psalm 90

THREESCORE YEARS and ten is the old English way of saying 70 (a score being 20). This reference in the Bible was adopted as a euphemism for a person's natural life, strange when you consider that it's only in the last few decades that average life expectancy has reached 70 (see **40**). The number threescore and ten appears a lot in the Bible.

■ 70 is a "weird" number. This means that its divisors, with the exception of itself, add up to more than itself, but no combination of any of its divisors adds up to itself.

$$1 + 2 + 5 + 7 + 10 + 14 + 35 = 74$$
$$But \ 35 + 14 + 10 + 7 + 5 = 71$$
$$And \ 35 + 14 + 10 + 7 + 2 + 1 = 69$$

There are no other weird numbers below 70. The next weird number is 836. Although it has been proven that there is an infinite number of weird numbers, nobody has yet found an odd one.

SR-71
also known as *Blackbird*, was a reconnaissance plane used by the US Air Force until 1998. On September 1 1974 the plane flew from New York to London in a record time of 1 hour 54 minutes and 56.4 seconds, about half the time it took Concorde.

★

❏ 71mph is the cruising speed of the fastest mammal on earth, the cheetah. Its acceleration has been measured as 0–45mph in two seconds, with top speeds in excess of 90mph. When hunting prey, a cheetah can maintain 71mph over 300 yards.

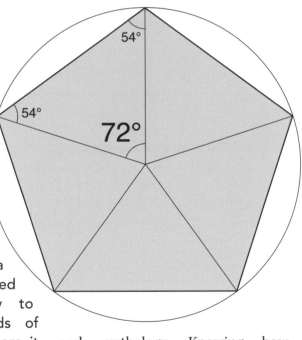

72 SEEMS TO HAVE a special meaning in several religions. According to Kabbalah (see **27**), there are 72 names of God. In Islamic tradition, 72 "wives" await the people of heaven. Jesus is said to have sent out 72 disciples.

But 72 also has some very definite applications in science. Perhaps these two things are not unconnected.

❏ 72° is a fifth of a circle. Therefore, it is a key measurement in drawing a regular pentagon, or a pentagram, the five-pointed star used in Christianity to represent the five wounds of Christ, and in Satanism, where it is inverted to represent anti-Christianity.

◼ 72 is divisible by every number from 1 to 10 apart from 5, 7 and 10, all three of which are factors of 70. Both 70 and 72 crop up regularly in religion and mythology. Knowing how mythology favors numbers like 6, 7, 10 and 12, it's easy to see how groups of groups will add up to 70 or 72.

❏ There are no fewer than 72 letters in the Cambodian alphabet, the longest alphabet in the world.

The Rule of 72 is a simple method used in financial circles to quickly estimate compound interest. If you want to know how long it will take to double your money, divide the interest rate into 72 and the answer will be the number of years. The margin of error increases as the interest rate rises: at 10 per cent interest it is accurate to a tenth of a year; at 20 per cent it is accurate to two-tenths.

The average human heart operates at 72 beats per minute.

72 IS THE MOST COMMON PAR FOR AN 18-HOLE GOLF COURSE.

72 hours equals three days.

72 in Computing

■ 72Hz is the lowest recommended refresh rate for a computer screen generally. This means that the screen image needs to refresh at least 72 times per second, otherwise it will appear to flicker to the human eye.

■ 72 dpi (dots per inch) is the standard resolution of a computer screen.

73

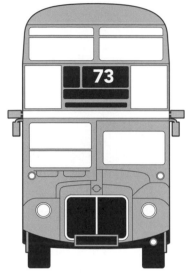

IF 666 IS the number of the beast, 73 is the number of the bus. Like 19, the 73 bus is a legend among London buses. It runs from Victoria to Tottenham and was claimed by comedians as their stereotypical bus route. The last Routemaster, the classic open-backed double-decker, was retired in 2004, but a modernized version was reintroduced in 2015 on the 73 and other routes. Talking of stereotypes, the expression, "the man on the Clapham omnibus," is shorthand for the average man in the street; it came from the 19th-century journalist Walter Bagehot who was looking for a way of summing up a hypothetical average Londoner though his invented passenger was bald and sat at the back of the bus.

★

*73 is a coded way of saying "all the best"
among amateur radio enthusiasts.*

There are five seasons of 73 days in the Discordian calendar. Discordianism is described as "a religion disguised as a joke disguised as a religion" and vice versa, and it is based on the acceptance that disharmony and chaos have a valid part to play in the universe. The five seasons align with the Gregorian calendar, starting on January 1, and are called:

Chaos	Confusion	The Aftermath
Discord	Bureaucracy	

74

A SEVENTY-FOUR was an 18th-century French warship that carried 74 guns. It was deemed a very practical design, offering firepower coupled with speed and manoeuvrability, and was used by other navies up until the steam age.

■ Seventy-five per cent of people use the phrase "seventy-five per cent" in preference to "three-quarters."

❝ *Many men die at twenty-five and aren't buried until they are seventy-five.* ❞

Benjamin Franklin, American statesman

ON DECEMBER 14 1998, contestant Ian Lygo made his 75th appearance on British gameshow *100%*, setting the world record for consecutive gameshow appearances. But this did not make him a hero. On the contrary, his winning run was brought to an abrupt end when the production company decided to change the rules. In response to complaints from disgruntled viewers, they imposed a new limit of 25 wins, and informed Lygo after his 73rd win that he could make two more appearances and no more. So, like Rocky Marciano, Lygo retired undefeated, with a bottle of champagne and winnings totalling a mere £7,500 (£100 for every show).

By contrast, on American gameshow *Jeopardy*, Ken Jennings matched Lygo's record in 2004, won $2.52million and became a legend, even though, unlike Lygo, he lost on his 75th appearance.

The Philadephia 76ers is an American basketball team that takes its name from the year of Independence (1776). The 76ers is the oldest franchise in the NBA, having formed in 1939 as the Syracuse Nationals. It won the NBA in 1955, 1967 and 1983.

✳

❝ *Seventy six trombones led the big parade,*
With a hundred and ten cornets close at hand.
They were followed by rows and rows,
Of the finest virtuosos,
The cream of every famous band ❞

"Seventy-six Trombones" by Art Meredith, which was the theme song from the 1957 musical *The Music Man*.

$$77 = 16 + 25 + 36 = 4_2 + 5_2 + 6_2$$

*In Sudoku, a puzzle may start with 77 filled-in squares
but still not have a unique solution.*

★

IN SWEDISH, 77 is a shibboleth. Put simply, this is Hebrew for a password, but it's more complicated than that. The term originates from a Bible story in which it was used to distinguish between two tribes with different dialects. Because of its complex pronunciation in Hebrew, anyone saying the word "shibboleth" could instantly be identified as belonging or not belonging to the tribe. In Swedish, 77 has a complicated pronunciation which could not be mastered by Norwegians or Germans and so instantly identified them as imposters.

■ American Airlines flight 77 was the third plane to crash on September 11 2001. The Boeing 757 was hijacked just before 9am and flown into the Pentagon in Washington DC, killing all 64 people on board and 125 in the Pentagon building.

77 Sunset Strip **was a trend-setting TV detective series of the late 1950s/early 1960s. It was the first hour-long private detective drama.**

**FORTRAN 77 WAS A COMPUTER PROGRAMMING
LANGUAGE DEVELOPED BY IBM IN THE 1950S.**

In some US states, dialing 77 on a cell phone will connect you to the police. This has led to an urban myth that 77 is a secret number that will get you the police in an emergency. In most cases, it doesn't work.

78

" 78 REVOLUTIONS A MINUTE "

THE ABOVE PHRASE, the title of a song by the Irish punk band Stiff Little Fingers, described the changing attitudes of the late 1970s, and was a pun on the 78rpm designation of gramophone records.

78s, as they were called, were the first record industry standard, set in 1925. Until then, records had been designed to rotate at a variety of speeds, usually somewhere between 74rpm and 82rpm. The 78 standard (actually 78.26rpm) was determined by the gear ratio in relation to the motor speed on the new electric turntable. Elsewhere in the world, where electricity supplies ran on different frequencies, the standard was 77.92rpm.

After World War II, record companies looked to develop discs to be played with a smaller stylus. Columbia Records launched the 33⅓rpm "long play" record (LP), and RCA Victor responded with its smaller, seven-inch 45rpm format, favored by jukebox manufacturers. 78s were produced into the 1970s, but mass production tailed off in the 1950s.

The disco movement of the 1970s made 12-inch singles popular, the larger format allowing extended versions of dance songs and great sound quality. By the 1980s the vinyl record was under threat from the compact disc. Because of their "noiseless" digital sound quality and their supposedly scratch-proof make-up, CDs soon surpassed LPs as the preferred format for albums, and by the 1990s they had taken over the singles market as well. However, connoisseurs still favor vinyl today, and indeed sales are on the increase.

■ 78 was the age of attorney Harry Whittington of Austin, Texas, when he was accidentally shot by Vice President Dick Cheney on February 11 2006. Cheney was aiming at a flock of quail and didn't see Whittington in his line of fire. Whittington received pellet wounds to the face and chest and suffered a minor heart attack, but recovered later in hospital. Cheney was cleared of any wrongdoing unlike his predecessor, Vice President Aaron Burr, who shot Alexander Hamilton in a duel in 1804. Burr was indicted for murder, but the case never came to trial.

79AD was the year Pompeii was destroyed by the eruption of Mount Vesuvius.

79 IS THE ATOMIC NUMBER OF GOLD.

AROUND THE WORLD IN 80 DAYS

IN JULES VERNE'S classic adventure, that is the number of days in which the hero, Phileas Fogg, bets that he can get around the world and back to the Reform Club in London. And he succeeds, though he doesn't realize at first.

Having elected to travel east, via among others Suez, India, Japan, America and Ireland, he arrives back in London despondent, having counted 80 days and five minutes "on the road." However, he hasn't taken into account the fact that by traveling east he has gained a day. When he is informed of this the next day, he manages to get to the Reform Club just in time to win the bet and collect his £20,000.

In February 2005, British yachtswoman Ellen MacArthur broke the world record for sailing single-handed round the world. Becoming the first woman to hold the record, she completed the 27,354 nautical mile journey aboard the trimaran *B&Q/Castorama* in 71 days, 14 hours.

The first non-stop flight around the world was by a team from the US Air Force in 1949. It took just 94 hours and 1 minute.

In 1974, Australian schoolteacher Dave Kunst completed the first verified walk around the world. Starting in Minnesota, USA, he flew across the Atlantic, Indian and Pacific oceans, but walked the rest of the way, a total of 14,450 miles in 1,568 days. Along the way he strode an estimated 20 million steps, went through 21 pairs of shoes and lost a brother, John, who was shot by bandits in Afghanistan.

The 80-20 Rule is a principle used in business circles. Also known as the Pareto Principle, it contends that in any situation, 80 per cent of the result is down to 20 per cent of the field. For example, a business will find that 80 per cent of its income comes from 20 per cent of its clients, and therefore it should concentrate on those 20 per cent, rather than wasting time on the others.

Vilfredo Pareto, after whom the principle was named, was an Italian economist who made the observation that 80 per cent of Italy's income was taken home by 20 per cent of the population. The 80-20 Rule itself was first suggested in 1937 by management guru Joseph M. Juran. As well as in business, the rule is said to apply in any walk of life, even horse racing!

■ The 1980s was a decade of major social and political change. In came a whole new vocabulary, out went an old order that seemed to have been around for ever.

IN: Solidarity   Mobile phone   Political correctness
Fatwa   CD   Yuppie   Personal Computer   AIDS   Crack
OUT: Iron Curtain   Berlin Wall   Apartheid   LP   Hippies   Rap

❑ 81 is a symbol of the Hell's Angels Motor-cycle Club, H being the eighth letter of the alphabet and A being the first.

There are 81 chapters in the ancient Chinese book the *Tao Te Ching.*

★

The lines in the palm of the left hand form the Arabic numerals for 81. The right palm forms 18.

★

82 IS THE ATOMIC number of lead and the number of games played in the NBA basketball league and NHL hockey league seasons.

139

A prime number and the sum of three consecutive primes, including twin primes:

23 + 29 + 31.

EIGHTY FOUR is a small town in Pennsylvania. The origin of its name is uncertain. It is not the only town in the world with a numerical name. There's a 6 in West Virginia, USA; a Twenty in Lincolnshire, England; a Treinta y Tres (33) in Uruguay; an Eighty Eight in Kentucky, USA; a Ninety Six in Carolina, USA; a Hundred in West Virginia; and a 1770 in Queensland, Australia. The latter was built on the site of Captain Cook's landing in May of that year.

IN APRIL 2006, Scott McDonald, chef at Selfridges department store in London, created a sandwich priced at £85 ($168). Weighing in at 21oz, the sandwich consisted of wagyu beef, fresh lobe foie gras, black truffle mayonnaise, brie de meaux, rocket, red pepper and mustard confit and English plum tomatoes, all encased in two slices of 24-hour fermented soda bread. The main expense was the wagyu beef, which comes from Japanese cows raised on a luxury diet of sake and massage. Selfridges' Food and Catering director, Ewan Ventner, said, "I think if you are a food lover, this represents great value for money."

EIGHTY-SIX is a code word among restaurant staff. Originally it meant that an item was no longer available, but it evolved to mean "get rid of something," also applying to unwanted customers. Its origin is unknown, although the most popular theory is that it was rhyming slang for *nix* (from the German *nichts*).

86 is the international dialling code for China.

★

There are 86 known metals in the periodic table.

87 is the sum of the squares of the first four primes:

$$2^2 + 3^2 + 5^2 + 7^2 =$$
$$4 + 9 + 25 + 49 = 87$$

And it is the sum of the divisors of the numbers 1 to 10:

1	1
1, 2	3
1, 3	4
1, 2, 4	7
1, 5	6
1, 2, 3, 6	12
1, 7	8
1, 2, 4, 8	15
1, 3, 9	13
1, 2, 5, 10	18
Total	**87**

WACO, TEXAS

A total of 87 people died in the Branch Davidian massacre at Waco, Texas, in April 1993. A raid by the Bureau of Alcohol, Tobacco, Firearms and Explosives led to a number of deaths and the FBI then laid siege to the cult's headquarters for 51 days, resulting in the high casualty list, which included children and cult leader David Koresh.

■ The Junkers 87 was a notorious German dive bomber of World War II.

87 is regarded as an unlucky number in some fields, particularly cricket, because it's 13 short of 100.

THE GETTYSBURG ADDRESS

"FOUR SCORE AND SEVEN years ago..." are the opening words of President Abraham Lincoln's famous speech at the dedication of the soldiers' cemetery at Gettysburg during the American Civil War. The War of Independence had been fought 87 years earlier.

The Gettysburg Address is a term often used to parody any long-winded oration, but it is a misnomer. There was a lengthy oration that day, delivered by the Honorable Edward Everett and lasting for two hours, but Lincoln's speech was short and sweet—disappointingly so for many of those present, who wanted more from their president and could not recognize the lasting resonance of his well-chosen words. Speaking for little more than two minutes, this is what Lincoln said:

"FOUR SCORE AND SEVEN years ago our fathers brought forth on this continent a new nation, conceived in liberty and dedicated to the proposition that all men are created equal.

Now we are engaged in a great civil war, testing whether that nation or any nation so conceived and so dedicated can long endure. We are met on a great battlefield of that war. We have come to dedicate a portion of that field as a final resting-place for those who here gave their lives that that nation might live. It is altogether fitting and proper that we should do this. But in a larger sense, we cannot dedicate, we cannot consecrate, we cannot hallow this ground. The brave men, living and dead who struggled here have consecrated it far above our poor power to add or detract. The world will little note nor long remember what we say here, but it can never forget what they did here.

It is for us the living rather to be dedicated here to the unfinished work which they who fought here have thus far so nobly advanced. It is rather for us to be here dedicated to the great task remaining before us—that from these honored dead we take increased devotion to that cause for which they gave the last full measure of devotion—that we here highly resolve that these dead shall not have died in vain, that this nation under God shall have a new birth of freedom, and that government of the people, by the people, for the people shall not perish from the earth."

THE International Astronomical Union has defined 88 constellations, divided into eight families:

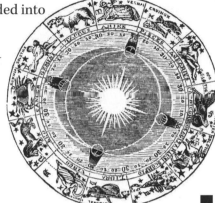

URSA MAJOR FAMILY
ZODIACAL FAMILY
PERSEUS FAMILY
HEAVENLY WATERS
ORION GROUP
BAYER GROUP
LA CILLE FAMILY
HERCULES FAMILY

In England, "Two Fat Ladies" is bingo slang for 88.

★

A piano has 88 keys.

■ 88 is a very lucky number in Chinese superstition because it means "doubly prosperous."

❏ The **Eighty-Eight** was a highly effective anti-aircraft gun used by the Germans in World War II. It fired an 88mm-caliber shell.

■ In the film *Back to the Future*, 88mph is the speed that must be reached in the professor's DeLorean car in order to travel back in time.

The Oldsmobile 88 was the marque's best selling car from 1950 to 1974.

89 IS A FIBONACCI prime number with interesting qualities. The reciprocal of 89 ($\frac{1}{89}$) gives a fraction with a sequence of 44 recurring digits:

0.01123595505617977528089887640449438202247191

What is remarkable is that this number is the sum of a sequence of numbers from the Fibonacci series, whereby each number moves a decimal point to the right.

0.0
0.01
0.001
0.0002
0.00003
0.000005
0.0000008
0.00000013
0.000000021
Etc., etc.

❝*Genius is 10 per cent inspiration and 90 per cent perspiration.***❞**

Thomas Edison, inventor of the light bulb

* 90 PER CENT OF STATISTICAL QUOTATIONS ARE MADE UP.

THE NINETIES

AS THE LAST DECADE of the millennium, the 1990s was a period of relative calm and political appeasement, as if the world was holding its breath and closing ranks. The peace processes in the Middle East and Northern Ireland both made significant moves forward, and the political map was altered by the rapid spread of democracy. With it spread capitalism, empowering many but widening the gap between haves and have-nots. It was the decade of globalization, in which computers truly took over daily life and the groundbreaking technology of the 1980s got more streamlined, compact and accessible. But at the same time, environmental and ethical issues triggered a growing movement towards basic values and being at one with nature.

'90S NOVELTIES	'90S NO MORES
The World Wide Web	Soviet Union
Email	East Germany
Extreme sports	Czechoslovakia
DVD	The European Community
Global warming	The franc, mark, peseta and lira
Human cloning	Princess Diana
DNA fingerprinting	Freddie Mercury
Genetically modified crops	Andres Escobar
Gay marriages	
Grunge	

An angle of 90° is called a right angle. Right angles are very important in construction, for making sure walls are "square," floors level etc. The Greek mathematician Pythagoras is famous for his theorem about the dimensions of right-angled triangles: the square of the hypotenuse equals the sum of the squares of the other two sides—the hypotenuse being the side opposite the right-angle. This theorem enabled builders of the time to construct perfect right-angles in their architecture. Knowing that sides in units of 3, 4 and 5 would make a right-angled triangle (since $3^2 + 4^2 = 5^2$), they were able to make a line marked with 12 equal segments and peg this out in a 3 x 4 x 5 triangle. The angle opposite the 5 segment side would always be a right-angle.

A tangent is a line that touches the circumference of a circle at 90° to the radius. This means it can only touch the circumference at one point, and then the two diverge, never to meet again. Hence the phrase "going off at a tagent," which means straying from the point in a conversation and never returning to it.

✳

Joe 90 was a nine-year-old special agent in puppet form, invented by Gerry Anderson (the creator of *Thunderbirds*) and broadcast on BBC television in the 1970s. Thanks to an ingenious computer called BIGRAT, Joe acquires the brain waves of his adoptive father Sam Loover, who works for the World Intelligence Network. Armed with this precocious knowledge, Joe becomes an agent for WIN, foiling baddies with the element of surprise.

91

91 is the sum of the squares of the numbers 1 to 6.

★

$$1_2 + 2_2 + 3_2 + 4_2 + 5_2 + 6_2 = 1 + 4 + 9 + 16 + 25 + 36 = 91$$
$$3_3 + 4_3 = 27 + 64 = 91$$

■ There are 91 days in the fiscal quarter that covers April, May and June. This varies from country to country as to whether it is the first financial quarter (as in the UK, Canada, and India for example), the third (as in the USA) or the last (as in Australia).

A SNUB DODECAHEDRON is a solid with 92 faces. This is called an Archimedean solid, of which there are 13. A Platonic solid (see **5**) is made up of one type of regular polygon, but an Archimedean solid uses two or more. The snub dodecahedron consists of 80 equilateral triangles and 12 pentagons.

■ **URANIUM**, with an atomic number of 92, is generally accepted as the heaviest of the elements that occur naturally on earth. However, the total number of naturally occurring elements is a subject of debate, due to the dwindling half-life of certain elements. Technetium (43) and Promethium (61) only remain in tiny traces on earth, as do Neptunium (93), Plutonium (94) and Californium (95). These are all regarded among the synthetic elements, leaving a current total of 90 naturally occurring elements.

UNITED AIRLINES FLIGHT 93 was the fourth plane to crash in the terrorist attacks in America on September 11 2001. The story of Flight 93 is the most intriguing of all from 9/11, because it remains unclear how the plane came to crash. Hijacked en route from Newark, New Jersey to San Francisco, California, the plane crashed into a field in Philadelphia, killing all those on board but not reaching any strategic terrorist target, unlike the three previously hijacked flights that day.

Flight 93 was delayed by 42 minutes, a crucial time lapse that enabled passengers on board to ascertain, via phone calls to friends and family, that the three atrocities had already occurred when their plane was hijacked. Therefore, they knew that there was no peaceful way out. Heroic passengers and crew fought back against the hijackers, but there is no evidence that they gained control of the cockpit, which would have been necessary to ground the plane.

One conspiracy theory claims that the plane was shot down by American fighters but there is no concrete evidence backing this up. The Commission investigating the incident concluded that the terrorists deliberately crashed the plane when they realized they were losing control of the situation.

"CONTINUED ON P94" is a familiar phrase to readers of *Private Eye*, the British satirical magazine, first published in 1961, that claims over 600,000 readers every fortnight. The random use of the number 94 is a running joke in the magazine. Many of its articles break off, with "continued on p94" printed beneath. Of course, there is no p94. *Private Eye*'s current editor, Ian Hislop, has been called the "most sued man in Britain."

★

94 is the atomic number of Plutonium.

❝Two per cent of the people think; three per cent of the people think they think; and ninety-five per cent of the people would rather die than think.❞

George Bernard Shaw, Irish playwright.

THE BIRTH OF PROTESTANTISM was initiated by a document drawn up by German monk **Martin Luther**, which became known as the **95 Theses**. In it, Luther challenged the Roman Catholic church on certain key doctrines and even questioned the authority of the Pope. He also queried the value of penance and the ethical justice of "indulgences," whereby people could buy time off from Purgatory by making generous donations to the church. When Rome demanded that Luther withdraw 41 of his 95 theses, he refused, and so sparked the **Reformation** which saw the church fragment.

96 is the number of people who died initially as a result of the **Hillsborough Tragedy** in 1989 (another died years later). The disaster occurred at Hillsborough Stadium in Sheffield, England, where Liverpool FC and Nottingham Forest FC were playing an FA Cup semi-final. With a huge crush of spectators trying to get into the ground, police opened a gate into a section of terracing that was already full, creating a further, fatal crush.

"My grandmother started walking five miles a day when she was sixty. She's ninety-seven now, and we don't know where the hell she is.**"**

Ellen de Generes, American comedienne

INTERSTATE 97 is the shortest Interstate in mainland USA. Built in 1993, I-97 runs from Annapolis to Baltimore, a distance of 17.62 miles, all within Anne Arundel County, Maryland.

97 HORSEPOWER IS THE SPEC OF THE LONDON OMNIBUS IN THE SONG BY FLANDERS AND SWANN.

"And will you succeed? Yes indeed, yes indeed! Ninety-eight and three-quarters percent guaranteed!**"**

Oh, the Places You'll Go by Dr Seuss

DR SEUSS (pronounced "Soice") was the pseudonym of American cartoonist and writer **Theodore Seuss Geisel** (1904–91). He was responsible for a large number of children's books, including *The Cat in the Hat* and *How The Grinch Stole Christmas*, both of which have been made into movies in recent times. He also worked for the American government during World War II, creating propaganda films and cartoons. *Oh, the Places You'll Go* was his last book, published in 1990.

"Ninety-eight per cent of American homes have TV sets, which means the people in the other two per cent have to generate their own sex and violence."

FRANKLIN P. JONES, AMERICAN BUSINESS GURU

98.6

98.6 °F (37 °C) is the "normal" body temperature of a healthy human being.

99

A 99 is a favorite ice cream in the UK. It consists of a cone of vanilla ice cream with a chocolate Flake in it. Why it is called a 99 is unclear. It certainly isn't to do with the price because when it first appeared it cost only a few pence. One theory is that Cadbury's, who make the Flakes, first gave them a batch number rather than a name because they were only made for sale to the ice cream trade.

★

One short of the magical 100 and a multiple of mythical 9 and the numerological 11, 99 is a number that trips lightly off the tongue, unless you happen to belong to the Basuto tribe of southern Africa, in which case you pronounce it:

"machoumearobilengmonoolemongametsoarobilengmonoolemong"

■ **ISLAM** talks about the *99 Names of God* and the prophet Mohammed is said to have declared that he who names them all shall enter paradise. The list of 99 names was compiled from careful study of the *Qu'ran* by a Muslim scholar called Al Waleed ibn Muslim, and is therefore held by some to be little more than man-made symbolism. However, some Muslims recite the 99 names in the full belief that they are the names enumerated by Mohammed.

❝*Floating in the summer sky 99 red balloons go by*❞

"99 Red Balloons"
(*99 Luftballons*) by Nena

This was a one-hit wonder for German band Nena in 1984. It told the story of a couple of kids releasing 99 red balloons into the sky, confusing the military into thinking there was some sort of invasion going on and triggering a nuclear holocaust.

100

> **❝***I think and think for months and years. Ninety-nine times, the conclusion is false. The hundredth time I am right.***❞**
>
> Albert Einstein

100 IS THE NUMBER of completion, as in 100 per cent, and a significant measure of time. A new century marks the beginning of a new era and the turnover point is often one of cultural change. The French expression *fin de siècle* (literally "end of the century") has connotations of decadence, coming as it does from the *Belle Epoque* at the turn of the 19th century, but it has assumed a broader sense of radical change and the death of the old order.

Hundreds appear frequently in the teachings and fables of all religions. 100 is a major milestone in sport and a popular number for list compilers, for example the Billboard Hot 100. The word "hundred" is derived from Germanic languages and may have originated from the German word "hund" (dog), being used as an approximation of the number of sheep a dog could keep in check. It ranged in value from 100 to 120. Later, a "hundred" came to mean an administrative measure of land containing 100 homes.

---

100 was regarded as divinely divine by Pythagoreans because it is the square of 10, "the divine decade." It is the sum of the first four cube numbers:

$$1^3 + 2^3 + 3^3 + 4^3 = 100$$

It is also the sum of the first ten odd numbers and the first nine prime numbers.

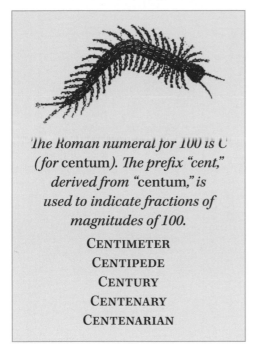

The Roman numeral for 100 is C (for centum*). The prefix "cent," derived from "centum," is used to indicate fractions of magnitudes of 100.*

CENTIMETER
CENTIPEDE
CENTURY
CENTENARY
CENTENARIAN

■ The prefix *hecto*, from the Greek *hekaton*, is also used in some instances to denote magnitudes of 100. For example, a hectogon is a 100-sided polygon, and a hectare is a measure of land equal to 100 ares, an are being a French word for an area of 100 square metres.

❑ A CENTENARIAN is a person who has lived to the age of 100. This feat of longevity is becoming more and more common, with numbers soaring exponentially in the last few years of the 20th century. A US census found that between 1990 and 2000, the number of centenarians in the USA alone had risen from 37,306 to 50,454. The worldwide centenarian population in 2000 was put at 180,000 and this has been forecast to rocket in the coming decades, according to a United Nations report.

★

❑ In the USA, a **CENTILLION** is a number with 100 groups of three noughts after the 1,000. In other words, 10^{303}. In Europe it is 10^{600}. A **googol** (see 10^{100}) is a 1 followed by 100 noughts.

★

"HUNDREDS AND THOUSANDS" ARE TINY PIECES OF CANDY USED FOR DECORATING CAKES. IN THE US THEY'RE KNOWN AS "SPRINKLES," WHILE IN HOLLAND THEY LIKE TO HAVE CHOCOLATE ONES SPRINKLED ON THEIR SANDWICHES.

★

■ A CENTURY is a score of 100 runs by one batsman in cricket. On April 14 2004, Brian Lara of the West Indies scored a quadruple century against England, setting a new record for the highest individual test score. Sachin Tendulkar of India tops the list of greatest number of test centuries scored with 51; Jacques Kallis of South Africa is in second place with 45.

AUSTRALASIA & OCEANIA
by numbers

Australasia is by far the least densely populated continent. In fact, there is an estimated one crocodile for every 220 people living there.

Area
2,968,000sq.miles/7,687,000sq.km

Percentage of global land mass
5%

Population (approx)   33,000,000

Population density (per sq.km)   4

Highest point
Carstensz Pyramid, Indonesia:
16,024ft/4,884m

Lowest point
Lake Eyre, Australia: 52ft/16m bsl

Longest river
Murray-Darling River: 2,310 miles/3,720km

Highest recorded temperature
53°C/128°F (Cloncurry, 1889)

Lowest recorded temperature
-23°C/-9.4°F (Charlotte Pass, NSW, 1994)

Estimated crocodile population
150,000+

101

T HE PALINDROMIC NUMBER 101 has found several prominent places in popular culture. It forms the unlikely connection between Walt Disney, George Orwell and British punk band, The Clash. *101 Dalmatians* was Disney's 1961 animated classic of Dodie Smith's 1912 book. Room 101 was Orwell's nightmare creation in his novel *1984*, a room containing the personal dread of anyone unfortunate enough to be put in there. In the case of Orwell's hero, Winston Smith, it was rats. Prior to joining The Clash, lead singer Joe Strummer fronted a band called the 101ers, not a reference to *1984* but the address of a London squat, 101 Walterton Road, where the band lived.

THE R101 WAS A BRITISH-BUILT AIRSHIP THAT CRASHED ON ITS MAIDEN VOYAGE OVER FRANCE, KILLING 48 PEOPLE ON OCTOBER 5 1930.

108

108 is a sacred number in Hindu and Buddhist culture. You'll find 108 beads on Buddhist rosaries; the Nepalese parliament has 108 seats; there are 108 Buddhist saints; 108 cowgirls were enamored of Krishna ... the list goes on. There are many reasons why 108 could be deemed a sacred number, some of them mystical, others mathematical.

108 is the product of two further sacred numbers: 9 x 12. It is also the sum of the squares of three sixes ($6^2 + 6^2 + 6^2$), but this would not have influenced the Hindus or the Buddhists.

■NUMBER 108's relationship to a pentagon and in turn to the Golden Ratio (see 1.618) is fascinating. The internal angles of a pentagon are 108° and the length of each side in proportion to the length of the sides of a five-pointed star drawn within the same parameters equals the Golden Ratio.

109

Soap bubbles always meet each other at an angle of 109° or 120°. This was discovered in 1873 by Joseph Plateau, a Belgian scientist.

110

BOTH OF THE TWIN TOWERS of the World Trade Center had 110 storeys, adding fuel to the numerological theory that 11 was a significant number in the 9/11 terrorist attacks. The Sears Tower in Chicago also has 110 floors.

✳

110 also links Osama bin Laden with Adolf Hitler, by way of Gematria (see **27**) and numerology. Osama bin Laden and Adolf Hitler both have a numerical value of 110.

A=1 **D**=4 **O**=15 **L**=12 **F**=6
H=8 **I**=9 **T**=20 **L**=12 **E**=5 **R**=18 =**110**

O=15 **S**=19 **A**=1 **M**=13 **A**=1   **B**=2 **I**=9 **N**=14
L=12 **A**=1 **D**=4 **E**=5 **N**=14 = **110**

These two men are also Birth Mates, by way of a practice that matches the numerical value of birth dates. Hitler's birth date is April 20 1889 (4/20/1889) and, if you keep adding the numbers until you're left with a single digit, you have the numerical value of 5: **4+2+0+1+8+8+9=32; 3+2=5**. Bin Laden was born on July 30 1957: **7+3+0+1+9+5+7=32**. By this method, they have further Birth Mates in Pol Pot and Lee Harvey Oswald, as well as Mick Jagger, Uma Thurman and Angelina Jolie.

110 is a camera film format, smaller than 35mm.

★

110 is the emergency number in Germany, Japan and China, among others.

★

110 PER CENT IS A PHRASE THAT HAS ENTERED POPULAR USE TO MEAN A SUPREME EFFORT. "WE GAVE 110 PER CENT": "WE GAVE EVERYTHING WE HAD AND MORE."

111 IS A NUMBER that readily attracts superstitions. This may or may not have something to do with the fact that it is the constant of a 6 by 6 magic square.

1	4	13	30	31	32
35	34	8	23	9	2
18	15	17	26	16	19
27	14	28	3	29	10
25	11	21	22	20	12
5	33	24	7	6	36

■ Also known as Nelson, 111 is considered an unlucky score in cricket. It may have got its name from the mistaken notion that Admiral Nelson had one eye, one arm and one leg. While the first two are true, Nelson was in full possession of both legs when he met his death at Trafalgar. When a batsman is on 111, it is traditional for his teammates sitting in the pavilion to keep their feet off the ground until he has safely moved on to a higher score. One particular umpire, David Shepherd, was known for hopping up and down until the score moved on.

There is some statistical evidence that 111 is an unlucky score, particularly for the Australians, who have been dismissed for that score on three key occasions in matches against England. The huge psychological pressure exerted on the batsman by this bizarre superstition is the most likely cause of them losing concentration at the vital moment.

Numerology

NUMEROLOGY is the study of the supernatural influence of numbers on human life. It grew up from the influence of ancient astrologers and mathematicians, coupled with religious mysticism and the occult. Today numerology seeks to find significance in the repeated occurence and patterns of numbers. Like 23rdians (see **23**), a lot of people claim to see the number 111 repeatedly in day-to-day life: on clocks, on car number plates, on computer screens etc. The numerologist's explanation is that this is a sign from the "spirit guides." Apparently they will whisper in your ear to get you to look at the clock just as it's saying 1:11, and what they're telling you is that you are thinking about starting a new cycle in your life and that your plans are sound.

Multiple digits like 111 are all significant for those who believe in spirit guides.

❏ 222 tells you to carry on doing what you're doing.

❏ 333 is a resounding yes to any questions you may be asking.

❏ 555 indicates that you have just had a thought or experienced an event that will change your life. Go with it.

❏ 666 tells you that your thoughts are unclear and you should not pursue them.

❏ 777 shows you that a valuable lesson has been learned.

❏ 888 is a warning to get ready for a new phase in your life.

❏ 999 tells you that what you're thinking marks the completion of a phase in your life.

❏ 000 means you are at one with the universe.

■ THE **F111** was a long-range strike bomber developed by the US Air Force and built by General Dynamics between 1967 and 1996. It was easily identified by its "swing wing" design which enabled it to take off and land at relatively low speeds with the wings forward, but achieve Mach 2.4 with the wings back.

112 IN IMPERIAL measures, 112lbs make a hundredweight. Twenty hundredweight make a ton. This correlation has been carried over to the metric system, a metric hundredweight being 50kg, one twentieth of a tonne.

The Qu'ran *is divided into 114 chapters called* suras. **114**

117 A CHINESE DRAGON traditionally has 117 scales. This is because of the significance of the number 9 in Chinese culture, where it is considered lucky. 81 scales are male and 36 are female.

$$81 = 9 \times 9$$
$$36 = 9 \times 4$$

Unlike in the West, where it is a fearful mythical beast, in Chinese culture the dragon, a combination of crocodile, snake and fish, is a sacred symbol of power and prosperity. As one of the 12 Chinese zodiacal signs, the Year of the Dragon is the most popular in which to give birth. The dragon became a symbol of the Chinese Emperor under the Xing dynasty and the number 9 was also imperial. Therefore, dragons and 9s have become intrinsically linked. There are nine Dragon Children, as well as nine types of dragon depicted in Chinese folklore:

CELESTIAL DRAGON   WINGED DRAGON
SPIRITUAL DRAGON   HORNED DRAGON
DRAGON OF HIDDEN   COILING DRAGON
TREASURES   YELLOW DRAGON
UNDERGROUND DRAGON   DRAGON KING

125

THE INTERCITY 125 was a high-speed train introduced in the UK in 1978. Recording a top speed of 148mph, it was the fastest diesel train in the world into the late 1980s. That record was topped on April 3 2007, by an electric-powered train, a French *Train à Grand Vitesse* (TGV), which reached 356mph.

★

125cc is one of the three classes of bikes used in Grand Prix motor racing. The other two are 250cc and MotoGP.

■ **128** is the sum of two squares ($8^2 + 8^2 = 64 + 64$), but it is the smallest number that cannot be expressed as the sum of two or more different squares.

128

139

"*I'm only too happy to be sitting absolutely motionless, doing nothing.*"

Queen Elizabeth II while sitting for her 139th royal portrait

The painting, commissioned for her 80th birthday, was described by the artist as "impressionistic." "I really like the way you get the blueness of the veins coming through the skin," he said.

144

144 is the square of 12. In other words, it's 12 dozen, which is also known as a gross. Twelve gross is called a great gross (1,728).

✳

The fact that 144 is a square number is surprising when you realize that it is also part of the Fibonacci Series. Because after 1, there are no other square numbers in the series other than 144. It also happens to be the 12th number in the series.

A decagon (10-sided polygon) has interior angles of 144°.

★

There are 12 dozen tiles in a mahjong set.

147

A 147 BREAK is the **perfect break** in snooker. It is the total score for potting all 15 reds (15 x 1), each followed by potting the black (15 x 7), and then clearing the colors: yellow (2), green (3), brown (4), blue (5), pink (6) and finally black again (7).

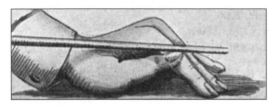

15 x 1 = 15
15 x 7 = 105
2 + 3 + 4 + 5 + 6 + 7 = 27
Total = 147

While this is regarded as the perfect break, it is possible to rack up a higher score. If a player commits a foul and leaves his opponent snookered, the opponent may pot any ball for a score of 1, followed by a color. If no other balls have been potted at this stage, he can still go on to achieve a perfect break, plus the free ball and color he began with. If that color was a black, he could achieve a maximum possible 155 points.

The first televised 147 break was by Cliff Thorburn in 1983 at the Embassy World Championship. It took so long it was nearly 1984 by the time he finished. By contrast, in 1997 Ronnie O'Sullivan completed a 147 break in just 5 minutes 20 seconds. In doing so, he earned himself a bonus prize of £165,000, which works out at £515.63 ($1,000) per second.

One hundred and eighty!

180

THAT'S THE CRY that greets any darts player who achieves a "maximum" with three darts by throwing three treble 20s. Darts players have phenomenal mathematical brains and can work out the score they require in split seconds. They're helped by the fact that there are a lot of landmark scores from which they know a familiar route down.

501—championship darts matches begin at this score. The classic route down to a finish is 60, 60, 60; 60, 60, 60; 60, 57, 24—or seven treble 20s, a treble 19 and a double 12. An alternative is to throw treble 18, treble 17, double 18 with the last three darts, and this was the route chose by John Lowe on October 13 1984, when he threw the first televised nine-dart finish. Throwing 167 (treble 20, treble 19, bullseye) with each three darts will also achieve a nine-dart finish, but this leaves no margin for error and is not seen in competition darts.

170—the highest position from which a player can finish. This requires a throw of two treble 20s and the bullseye (50).

163—the lowest score that is impossible to make with three scoring darts. The others are 166, 169, 172, 173, 175, 176, 178, 179 and anything above 180.

110—the highest finish obtainable with two darts.

40—known as "double top," the double 20 is one of the favorite out shots.

32—double 16 is also a favorite, because 16 is next to 8. If you miss the double and just score a single 16, you only have to adjust your aim slightly for the double 8.

26—one dart in the 20 and one to each side, in the 5 and the 1. Known as "bed and breakfast" or "a classic."

3—known as "the madhouse" because it requires a finish of 1 and double 1, leaving nowhere to go if you land in another bed.

180 The difference between the point at which water freezes and the point at which it boils is 180° on the Fahrenheit scale. It freezes at 32°F and boils at 212°F.

★

TO "DO A 180" IS TO TURN AND FACE IN THE OPPOSITE DIRECTION—A TURN THROUGH 180°. THE INTERNAL ANGLES OF A TRIANGLE ALSO ADD UP TO 180°.

200 IS ONE OF THE most common ISO numbers of camera film. ISO stands for the International Organization for Standardization and it rates film for its sensitivity to light. The higher the number, the more sensitive and, therefore, the less exposure is required.

200 ISO is in the middle of the scale of the most common film types, which range from 25 up to 3200. In the shops you'll generally find 100, 200 and 400 because these require less technical knowhow to achieve good results.

The difference from one film speed to the next corresponds to one f-stop on a camera. So, with the same shutter speed, a shot taken at an aperture of f8 with 100 ISO film will have the same exposure as one taken at f11 with 200 ISO. Likewise, if you keep the aperture the same but double the shutter speed for the 200 film, you'll get a similar result.

The results are not exactly the same because the aperture and film speed influence other qualities in the photograph, such as sharpness and depth of field. The higher the ISO rating of film, the more grainy the photograph. This is used by professional photographers to give a certain effect, but for the amateur happy snapper, too much graininess is usually not good.

200m is the second-fastest sprint distance in athletics. It is also the longest distance raced in swimming for every stroke other than freestyle. 200m is four lengths of a competition pool.

On July 27 1963, Don Schollander of the USA became the first man to swim 200m freestyle in under two minutes. His time was 1.58.8. By the end of the 20th century, the record stood at 1.46.00, an impressive 36 seconds faster than at the start of the century.

216 IS A NUMBER that has been given widespread spiritual significance. Ancient mathematicians regarded it as special because it is the product of the first two cubes (discounting 1). It is the cube of 6, the sum of the cubes of 3, 4 and 5 and also the product of the cubes of 2 and 3.

$$2^3 = 8$$
$$3^3 = 27$$
$$8 \times 27 = 216$$

In Christianity, 666 is a highly significant number and some see 216 as a representation of this, since 6 x 6 x 6 = 216.

It is the constant in a 3 by 3 multiplying magical square. Each row, column and diagonal multiplies to 216.

2	**9**	**12**
36	**6**	**1**
3	**4**	**18**

■ In the Hebrew Kaballah, a 216-digit number is believed to represent the true name of God. This features in the 1998 movie π (Pi).

＊

❏ 216 is also the smallest cube that is the sum of three cubes:

$$3^3 + 4^3 + 5^3 =$$
$$27 + 64 + 125 = 216$$

220 BOASTS THE UNIQUE record of being the lowest amicable number. Its partner in this amicable arrangement is 284. Amicable numbers are pairs of numbers whose factors (except the number itself) add up to the other number.

The factors of 220 are 1, 2, 4, 5, 10, 11, 20, 22, 44, 55, 110. Add these together and you get 284.

The factors of 284 are 1, 2, 4, 71 and 142. Add these together and you get 220.

Legend tells of a sultan with a penchant for puzzles. Learning that he had a mathematician imprisoned in his jail, he decided to strike a deal. The sultan told the mathematician that if he gave him a problem to solve he could go free for such time as the sultan took to solve it. Once the sultan had solved the problem, however, the mathematician would be beheaded.

So the mathematician told the sultan about the amicability of 220 and 284 and challenged him to find another pair of amicable numbers. The sultan never managed it and the prisoner died of old age, a free man.

Fortunately for the prisoner, the sultan didn't have the services of Leonard Euler, the 18th-century Swiss mathematician who found 59 pairs of amicable numbers.

The amicability of 220 and 284 was known to Pythagoras, who held them in high esteem. The numbers were regarded as having special power and were used to represent amicability in various forms of symbolism. But it took a while for anyone to find more. The next pair are 1,184 and 1,210, but these weren't discovered until 1866, when a 16-year-old Italian called Paganini revealed them to the world. Euler and other great mathematicians like Descartes and Fermat had missed this pair, as had the Arab mathematician al-Banna, who found the second pair to be discovered, 17,296 and 18,416, around the turn of the 13th century.

It is believed that all the amicable numbers below 1,010 have been discovered, but it has not yet been determined whether or not there is an infinite number of pairs. Neither has anyone found, or proved, that it is impossible to have, a pair consisting of one odd and one even number.

Here are the first few that we know:

220	284
1,184	1,210
6,232	6,368
10,744	10,856
17,296	18,416
9,363,584	9,437,056

256

256 IS THE TOTAL number of games played in a season of the NFL and it is the number of characters in the 8-dot Braille system that replaced the original 6-dot system.

■ 256Hz is the frequency of middle C beloved of many classical composers. This frequency is known as Philosopher's or Scientific middle C because it fitted in a scale regarded by the likes of Pythagoras as most pleasing. This scale was formed using simple ratios, Pythagoras believing that simple numbers would be more pleasing to the ear. Not knowing about sound waves and frequencies, Pythagoras effected his scale by adjusting the lengths of the strings, using the following ratios:

Octave: 2 to 1; Fifth: 3 to 2; Third: 5 to 4

However, there was a problem with this system. It was not mathematically consistent across all the keys, and therefore switching between certain keys produced ugly tonal clashes. So people experimented with different tuning systems which gave a more pleasing result. Johann Sebastian Bach's "well tempered" scale was a major improvement, as he proved by writing a piece using all 48 keys called *The Well Tempered Clavier* (see 48).

This inspired the development of the even tempered scale, in which every semi-tone was the same interval. This brought much more complex frequencies into play, and it led to numerous disputes over what the ideal concert pitch should be.

Today, concert pitch is based on a frequency of 440Hz for the A above middle C, which puts middle C itself at 261.6Hz. However, there is a strong movement campaigning to return to the classical 256Hz at which so many musical works were written, largely because singers are struggling to perform at the higher pitch.

Here are some more familiar frequencies:
20,000Hz—highest pitch audible to the human ear
4186.01Hz—top C on a piano
4000Hz—the most irritating frequency known to man
432Hz—first cry of a new-born baby
350Hz—the average human voice
256Hz—Philosopher's middle C
111Hz—conducive to trance
53Hz—the beat of a hummingbird's wings
40Hz—thunder
25Hz—cat's purr
20Hz—lowest pitch audible to the human ear
3Hz—low voice of an elephant

270

270m (891ft) is the height of the **Millau Viaduct**, which broke the record as the world's tallest road bridge when it opened on December 14 2004. The bridge spans the valley of the Tarn, which runs through the Massif Central in France. It was built to enable faster travel between Paris and the South of France, took three years to construct and cost €394 million. It is 1.6 miles long and suspended on seven pillars, the tallest of which is 342m high (1,122ft), taller than the Eiffel Tower. Drivers crossing the Millau Viaduct often find themselves above the clouds, and liken the experience to flying.

THERE ARE 360° in a circle, and in the internal angles of a quadrilateral (four-sided polygon). The world is divided into 360 degrees of longitude, 180° east and 180° west.

360

The idea of dividing a circle into 360° belonged to the Sumerians, who lived in the Middle East around 5,000BC. They were expert astrologers and mathematicians and they were also one of the first civilizations to use writing, so we have evidence today of their mathematical knowledge.

The Sumerians used a base 60 counting system (see **60**) and they also discovered the close relationship between a circle and a hexagon (six-sided polygon). They knew that each side of a regular hexagon was equal to the circle that circumscribed it (see diagram), and therefore it was natural to break the circle down into six segments. Each of these segments they then subdivided into 60 smaller segments, giving a total of 360 degrees, as we now call them.

They then used this figure, and its simple sub-divisions of 30 and 12, to divide the night sky. They also had a calendar year of 360 days. 360 is the smallest number with 24 factors, making it extremely convenient to divide up into months, days, minutes, seconds etc.

365.25

THE ACTUAL NUMBER of days in a **solar year** according to the Julian calendar. A solar year is the time it takes for the Earth to make one complete rotation of the Sun. The exact figure is actually more like 365.2421895134, but this was rounded up to 365 and a quarter by Julius Caesar, who introduced a leap year every four years to accommodate the four extra quarter days. However, because it is not exactly a quarter day, the leap year is not quite

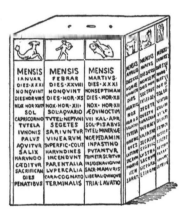

enough to keep the calendar exactly synchronized with the solar year. Therefore, further adjustments have to be made. The Gregorian calendar set the solar year at 365.2425 days and established that every 100th year should not be a leap year, even though it is a multiple of four; and every 400th year, even though it is a multiple of 100, should be a leap year. This keeps things pretty much in check, even though the solar year shortens by 0.00000006162 days every year.

Whatever the exact figures, the solar year is probably the most significant measure of time, each year having its own unique number, which then becomes associated with historical events. For example:

SPECIFIC YEARS HAVE INSPIRED WRITERS OF MUSIC AND FICTION:

1812—OVERTURE BY TCHAIKOVSKY

1977—SONG BY THE CLASH

1984—NOVEL BY GEORGE ORWELL

1999—SONG BY PRINCE

2001—NOVEL AND FILM BY ARTHUR C. CLARKE AND STANLEY KUBRICK

2525—SONG BY ZAGER AND EVANS

Note, the last four were all futuristic, written prior to the year in question.

374

THE OLDEST RECORDED animal on earth, the *Arctica islandica* **Icelandic Cyprine mollusc**, was 374 years old. This mollusc lives in colder Atlantic waters, close to shore, and grows to about 10cm across, although growth is extremely slow.

Galapagos tortoises are believed to live beyond 200 years, as are some species of whale. Tui Malilia, a tortoise given to the Tongan

royal family by Captain Cook in 1777, lived until 1965, a confirmed total of 188 years.

A macaw named Charlie, living in England, is reported to have been alive since 1899. However, claims by the owner that it once belonged to Winston Churchill and still utters anti-Nazi obscenities taught to him by the late British prime minister are denied by the Churchill family.

In October 1995, Jeanne Louise Calment of France was proclaimed as the oldest recorded human when, at the age of 120 years, she surpassed the record held by Shigechiyo Izumi of Japan. Calment was born on February 21 1875 and died on August 4 1997 at the age of 122 years, 5 months and 14 days.

432 IS regarded in some circles as a significant factor in a number of religions and mythical traditions.

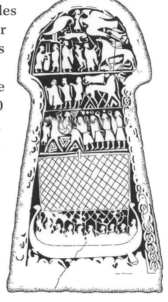

The Kali Yuga, or dark age, is a period of time in Hindu tradition, said to amount to 432,000 years, during which man's morality degenerates.

In Babylonian legend, there is said to be a dynasty of ten kings who reigned for a total of 432,000 years up to the time of the great flood.

In Norse mythology, the Doomsday of the Gods will feature 800 warriors coming through the 540 doors of Valhalla: a total of 432,000.

Astrologers calculate that the time required for the equinox to travel through one degree is 72 years, and therefore to travel through all Zodiacal houses takes 25,920. (In actual fact it's quicker than this, but it speeds up every year, so this would have been a fairly accurate calculation in the time of the Sumerians). The Sumerians counted in base 60. 25,920/60 = 432.

★

*F*AHRENHEIT 451 is a book by Ray Bradbury set in the future during a regime of heavy censorship in which books are banned. The central figure, Guy Montag, has the job of a "fireman," which means he burns books. This is where the title comes from: 451°C (233°C) being the temperature at which paper burns.

This is actually the auto-ignition point, the temperature at which a substance catches fire without contact with a flame or spark. As you might expect, paper has a low ignition point compared to other fuels. Bradbury's novel was made into a film in 1966 by François Truffaut, starring Oskar Werner and Julie Christie. The title inspired Michael Moore's 2004 documentary film *Fahrenheit 9/11*.

THE OCEANS
by numbers

71% of the Earth's surface
97% of the Earth's water

Total volume of water
1,347,000,000cu.km/322,280,000cu.
miles (296,298,447,105,777,900,000
gallons)

Average surface temperature
17°C/62.6°F

Average deep ocean temperature
0–3°C/32–37.5°F

Ocean	Area (miles²)	Deepest point
Pacific	64,000,000	36,198ft (Mariana Trench)
Atlantic	33,000,000	28,231ft (Puerto Rico Trench)
Indian	28,000,000	25,345ft (Java Trench)
Southern	7,900,000	23,736ft (South Sandwich Trench)
Arctic	5,000,000	17,881ft (Eurasia Basin)

500

THE INDIANAPOLIS 500 and Daytona 500 are two famous races in the American IndyCar and NASCAR series respectively. In both cases the 500 stands for the 500-mile distance of the race.

The Indianapolis 500 is one of the oldest motor races in the world, dating back to 1911. It is a notoriously dangerous race, which has claimed the lives of 38 drivers, 16 mechanics and ten spectators.

The Fiat 500 was one of the world's first small city cars, built between 1957 and 1979. The joke was that it was powered by a rubber band.

*

On March 9, 2019, speed skater Pavel Kulizhnikov or Russia set a 500m world record time of 33.61 seconds, at the Winter Olympics.

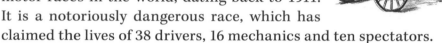

501 IS THE NUMBER of one of the world's most popular brands of jeans, made by Levi's. It should be noted that Levi Strauss, the Bavarian immigrant who pioneered the manufacture of blue jeans in 1873, is not the same person as Claude Lévi-Strauss, the French anthropologist.

★

501 is the starting score in competition darts.

★

❏ 501 is also the highest individual score in first-class cricket, racked up by Brian Lara off 427 balls for Warwickshire against Durham in 1994.

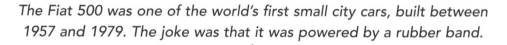

555 ◼ The **fake telephone exchange** number used in the movies. Because of the growing demand, many numbers beginning 555 have had to be allocated to real users. But the numbers 555-0100 up to 555-0199 are still unassigned.

666

666 IS THE MOST sinister number of all but it highlights the fact that, while the world is made up of many religions, not all of them can be right. For while 666 is the Devil's number in Christianity and symbolizes the coming of the end of the world, in Chinese culture it sounds like "things going smoothly" and is regarded as very lucky. Two very different interpretations of the same intriguing number.

★

Some Greek versions have the number at "six-hundred ten six" (616). Whatever way you interpret this, it does seem to be saying that here is a number that needs to be examined. And for hundreds of years, scholars have duly obliged (see page 174).

★

■ THE ORIGIN of the diabolical nature of 666 lies in the Book of Revelation 13:18. Originally written in Hebrew and translated into Greek, this passage has been interpreted in many ways, but a fairly reliable translation is found in a 1982 hit single:

❝Let him who hath understanding reckon
the number of the beast
For it is a human number, its number
is six hundred and sixty six**❞**

"The Number of the Beast" by Iron Maiden

MOST POPULAR 666 THEORIES

The non-specific theory

DCLXVI IS 666 in Roman numerals. And DCXVI is 616. Both, it is claimed, were used as expressions of unspecific large numbers, as we might say "thousands" or "millions." This interpretation suggests that there is indeed a literal beast with a very high number.

The Caesar theory

ANOTHER INTERPRETATION is that the number is code for a Roman who was regarded by the Christians of the time as a major threat. The Emperor Nero is a popular candidate for this theory, as his name in Greek has a numerical value of 666. Another suggestion is that it referred to Divus Claudius, whose mother even thought he looked like a beast. His initials DC give us 600 in Roman numerals; the Greek word for Caesar has a Gematria value of 60; and he was the sixth emperor after Julius Caesar: six-hundred sixty six. So could this version simply have been having a dig at the Romans?

The imperfection theory

JUST AS 7 is seen as the number of perfection, God's number (see **7**), and 8 as the number of Jesus, 6 is regarded as the number of imperfection: man. Man was created on the sixth day. Being imperfect in mind, body and soul, it is argued that 666 is simply man. "For it is a human number."

The numerology theory

NO NUMEROLOGIST worth their salt could pass up such an invitation to find the hidden meaning in a number like this, and countless forms of Gematria (see **27**) have been used to throw up the name of the Beast. Some use a straightforward numerical conversion: A=1, B=2 etc. Others use a similar conversion, but raise the sequence beyond J=10 to K=20, L=30 etc. One sequence uses the six times table: A=6, B=12 etc; another reverses this: Z=6, Y=12 etc. The Hebrew and Greek Gematria are also applied in many cases and the results of all these theories are quite staggering. If you mix the numbers up enough, almost anyone will fit. Here's a list of some of the more prominent figures that have been fingered as a result:

ADOLF HITLER	JOHN F KENNEDY	SANTA CLAUS
BILL GATES	POPE JOHN PAUL II	THE USA
MARTIN LUTHER	JOSEF STALIN	THE COMPUTER
NAPOLEON BONAPARTE	BORIS YELTSIN	THE WORLD WIDE WEB
RONALD REAGAN	SADAM HUSSEIN	*THE HOLY BIBLE*
HENRY KISSINGER	AYATOLLAH KHOMEINI	

The anti-capitalist theory

"THE LOVE OF MONEY is the root of all evil," is another biblical phrase (Timothy 6:10). The spread of capitalism is seen by many as the greatest evil in modern times, and so they've set to work to find how 666 may be relevant to money. Lo and behold, 666 is the total value of the numbers on a roulette wheel. But it goes further. Look at a bar code on your groceries and you should see three pairs of narrow lines, one at each end and one in the middle. These pairs of lines are identical in appearance to the symbol for the number 6. In fact, to the scanner they are not read as 6, but to the untrained eye they are perceived as 6. Therefore, the number 666 is intrinsic to international commerce. Furthermore, the quest for ever-greater security is leading towards doing away with credit cards and carrying our bank details in a code tattooed on our hand. The Book of Revelation says that every man shall carry the mark of the beast? Was this a reference to capitalism?

■ THE 1976 movie, **The Omen**, launched three years after the bar code, was responsible for a huge upsurge in interest in the number 666. The film and its sequels portrayed the coming of the antichrist in the form of a boy called Damien, who had a 666 mark on his head and caused appalling "accidents" to befall anyone who stood in his path. The film was remade 30 years later with no improvement and released on June 6 2006 (6.6.06).

761

THE SPEED OF SOUND varies according to the medium it is traveling through, which means temperature and altitude have a major bearing. For example, at 80,000 feet sound travels at 660mph.

On August 16 1960, US Air Force Capt. **Joseph W Kittinger, Jr** achieved a speed of 714mph in freefall, making him the only human to break the sound barrier without being inside a machine. As part of an Air Force research programme called Project Excelsior, which was looking into the use of parachutes, he was dressed in an astronaut suit and lifted to 102,800 feet on a platform attached to a balloon. Stepping off over New Mexico, he plummeted 84,700 feet in 4 minutes, 36 seconds, reaching 714 mph. After he opened his parachute, it took him a further 9 minutes, 9 seconds to reach the ground, where he landed safely.

On September 25 1997, Briton Andy Green became the first man to break the sound barrier on land, achieving a speed of 763.434mph over 1km of the Black Rock Desert, USA.

The first man to fly a plane through the sound barrier was American pilot, Chuck Yeager, who achieved the feat in a Bell XS-1 on October 14 1947, at an altitude of 45,000 feet.

Everybody who can crack a whip is capable of breaking the sound barrier. The crack sound is caused by the tip of the whip traveling faster than the speed of sound, and thus causing a small sonic boom.

Just as 666 is considered the embodiment of evil, 777 is regarded by some Christians as a representation of the Holy Trinity. It's certainly good news if you're playing the slot machines in Las Vegas, as a line of three sevens means payout time.

■ The Christian allusion to 777 was turned on its head by notorious satanist **Aleister Crowley** (1875–1947) when he published an occultist's handbook entitled *Liber 777*. This book is still a guiding light for those who like to indulge in satanism and witchcraft.

Crowley himself was the son of Plymouth Brethren. Born in Leamington Spa, in the heart of England, he threw off his religious upbringing while at Cambridge University, became a heroin addict and set about earning a reputation for witchcraft and perverse sexual practices. He proclaimed himself to be the Antichrist and adopted the number 666 as his own. At the height of his powers he was branded "the wickedest man in the world."

❏ The Boeing 777-200LR Worldliner is a pioneering jet airliner with a range of almost 11,000 miles fully laden. Its introduction in 2005 hailed a breakthrough for long-haul air passengers because it meant they could fly between virtually any two cities in the world without stopping.

In 2005 it set a new record for a long-distance commercial flight of just over 13,500 miles, between Hong Kong and London. The flight took 22 hours and 22 minutes. In 1903, Wilbur Wright took 12 seconds to fly 128 feet. Minutes later his brother Orville stayed airborne for 59 seconds, covering 915 feet.

900

900 is the prefix of premium (pay per call) telephone numbers in the USA and its enclaves.

911

911 IS THE emergency number in the North American telephone system. It's paradoxical that it should also be the date of America's worst disaster on home soil, the terrorist attacks of September 11 2001.

Every year on 9/11, a commemorative service is held at the site where the twin towers of the World Trade Center stood in Manhattan, and the names of the victims are read out. During the reading of the names they pause four times, marking the times that the two planes struck the towers (8.46am and 9.03am) and the times at which the towers collapsed (9.59am and 10.29am).

Many people try to find secret meanings in the numbers involved in 9/11, but the most powerful significance attaches to the casualty list, the sheer number of those who died:

THE DEAD

- ❏ American Airlines Flight 11 . . . . 88 + 5 hijackers
- ❏ United Airlines Flight 175 . . . . 59 + 5 hijackers
- ❏ American Airlines Flight 77 . . . . 59 + 5 hijackers
- ❏ United Airlines Flight 93 . . . . 40 + 4 hijackers
- ❏ World Trade Center 2,602
  (24 still listed as missing)
- ❏ The Pentagon 125
- ❏ Total 2,992

OF WHICH

- ❏ NYC Fire Department 343
- ❏ NYC Police Department 23
- ❏ Port Authority Police Department 37
- ❏ Children under 11 8

THE DAMAGE

- ❏ 9 buildings destroyed or later demolished in Manhattan
- ❏ 4 planes destroyed
- ❏ 4 floors of the outer ring of the Pentagon, about 35m wide, collapsed, destroyed by fire

THE WORLD'S FIRST dedicated emergency telephone service was the **999 service** implemented in London in 1937. It was introduced following an inquiry into a fatal fire, which had concluded that lives could have been saved had the emergency calls had a faster route to the necessary services. Up until that point, emergency calls were routed by an operator to the police Information Room on Whitehall 1212.

The number 999 was chosen for several reasons. It could not be dialed in error by a faulty telephone, as 111 could; its use did not require the closure of any existing telephone exchange; and public telephones could easily be adapted to take 999 calls free of charge. Today, however, 999 is perhaps too easy to dial on modern push-button phones. Around 75 per cent of 999 calls are not actual emergencies. Many of these are misdials, largely caused by children playing with the phone. A new number, 101, is being introduced to handle incident reports that are not emergencies, and thus take the burden off 999.

999 was the number of the US locomotive, *Empire State Express*, the first train to exceed 100mph.

★

999 IS 666 ROTATED THROUGH 180°.

■ **999-year** leases are commonplace in property transactions. This extended lease term is thought to have originated as a loophole in a law passed in England outlawing 1,000-year leases because a lease term that long was effectively a sale. There is also a significant psychological threshold between 999 and 1,000, just as $9.99 seems so much cheaper than $10.00.

1,000

THE HIGHEST NUMBER represented by a single Roman numeral (M short for *mille*) and the first number to take a comma, 1,000 is a major milestone in the modern counting and measuring system. The Latin *mille* remains in evidence, as does the Greek *khilioi* (kilo).

MILLIMETRE	A THOUSANDTH OF A METRE
MILLIGRAM	A THOUSANDTH OF A GRAM
MILLENNIUM	A THOUSAND YEARS
MILLIPEDE	CREATURE WITH A THOUSAND LEGS
KILOWATT	A THOUSAND WATTS
KILOMETRE	A THOUSAND METRES
KILOGRAM	A THOUSANDS GRAMS

We also get the word "mile" from *mille*, being a thousand paces in Roman measures. And the prefix "kilo" is commonly abbreviated to "k" and applied to money ($10k = $10,000).

✳ **THE END** of a millennium is seen as a time of great upheaval, even in these modern times. The year 2000 marked the end of the second millennium in the modern calendar, and approximately the 200th since the evolution of man. Just as the end of the first millennium AD induced mild panic in some quarters, with farmers not bothering to grow crops because they thought the world was about to end, the planet was gripped by fear of an apocalypse, in the form of the Y2K bug. Basically, this was a programming anomaly that meant some computers were not set up to recognize dates after midnight on December 31 1999, and therefore, it was feared, could throw the world's electronic systems into turmoil—or at least ensure your video failed to record episode 4,000 of *Friends*. As a result, an estimated $1 trillion dollars was spent on IT solutions worldwide. The IT industry had never had it so good. Unfortunately for them, such bonanzas only come along once every thousand years.

THOUSAND ISLAND dressing gets its name from an area on the border of Canada and the USA, where the St Lawrence river enters Lake Ontario. The name is only a rough approximation; there are actually 1,864 islands and because some of these are tiny there are rules to decide what actually constitutes an island. 1, they must be above water every day of the year; 2. They must have at least two trees growing on them. Thousand Island dressing is a blend of mayonnaise and ketchup with chopped pickles, red bell peppers, olives and hardboiled egg. It is sometimes given a kick by Worcestershire sauce or chilli. It was given its name by an actress named May Irwin, who tasted it while on a fishing expedition in the area. The fishing guide, George Lalonde Jr, would feed it to his customers with the fish they caught, and it was made by his wife, Sophie. May Irwin requested the recipe and passed it on to a friend of hers who owned the local Herald Hotel (now the Thousand Islands Hotel). She also gave it to George C. Boldt, who owned the Waldorf Astoria in New York, and it was his restaurant manager, **Oscar Tschirky**, who is credited with introducing it to the world.

1,001

A THOUSAND AND ONE is a phrase used to express an innumerable amount of something. "I've told you a thousand and one times not to do that!"

A Thousand and One Arabian Nights (originally *A Thousand Nights and a Night*) is a collection of stories concerning the cunning Queen Scheherazade of Persia, who comes up with an ingeniously creative way of preserving her life. Her new husband, King Shahryar, is a notorious misogynist who is in the habit of executing his new brides after just one night because he suspects all women of being fundamentally unfaithful. In order to avoid suffering the same fate, Scheherazade embarks upon telling him a series of bedtime stories, each with a cliffhanger ending, so that he will have to keep her alive to hear how the story continues. By the time she has told 1,001 stories (a period of 143 weeks), she is the proud mother of three sons and Shahryar grants her a pardon.

1776

❝We hold these truths to be self-evident, that all men are created equal, that they are endowed by their Creator with certain unalienable Rights, that among these are Life, Liberty and the pursuit of Happiness...❞

1776 IS A NUMBER etched on the heart of every patriotic American, for that was the year of the **Declaration of Independence**. The declaration was drafted by Thomas Jefferson on July 4 and signed a month later by every member of the Second Continental Congress, the collective body of governors in the 13 states, acting in defiance of British rule. Independence had actually been declared on July 2 in a statement written by Richard Henry Lee, but Jefferson's rewrite was formally adopted by Congress on July 4 and that day stands as Independence Day. It took a further seven years for the British to surrender control at the Treaty of Paris, having been outgunned in the Revolutionary War by an alliance of France, Spain and Holland.

■ TODAY, 1776 is commemorated in a number of ways. Several airlines run a flight 1776 to Philadelphia and, on the 200th anniversary of the Declaration, Pan Am changed the number of its flight from New York to London from 100 to 1776.

❝*It was a bright cold day in April, and the clocks were striking thirteen.*❞

1984

THIS SINISTER SENTENCE introduces us to another unforgettable year, 1984, which, thanks to the imagination of **George Orwell**, remains a synonym for futuristic discord and totalitarianism, despite the fact that the year itself is rapidly fading into history. Orwell, who had already made a name for himself with a number of critically acclaimed novels, including *Down and Out in Paris and London* and *Animal Farm*, embarked upon his ninth book in 1948, the eventual title being a simple rearrangement of that date. Its bleak portrayal of totalitarianism earmarked *1984* as a dangerously political novel and it was banned in many countries that saw themselves as the target of its theme. But since its publication in 1949 it has been translated into over 60 languages and continues to sell in vast numbers. Orwell didn't realize it at the time, but *1984* would be his last novel. He died of tuberculosis on January 21 1950.

When 1984 came round, interest in the book saw sales soar. Over 300,000 copies were sold in the UK alone, 35 years after publication. Elsewhere, there were apocalyptic overtones in the air, particularly in India where the Bhopal disaster and Indira Gandhi's assassination took place.

4,844

4,844 was the estimated age of the world's oldest recorded plant, a Great Basin bristlecone pine (*pinus longaeva*) called **Prometheus**, when it was felled in Nevada, USA, in 1964. The ring count of 4,844 was taken at a section several feet above the growth point, so it's possible that the tree may have been over 5,000 years old. Prometheus was cut down by scientists who were not aware at the time of its record age. That left Methuselah, another bristlecone pine growing in California, as the oldest living plant on record, at an estimated 4,800 years old.

This is not, however, the oldest living organism on earth. Some plants, such as vines, fungi and berry bushes, grow in clonal colonies, which means new shoots come and go from the same original spore, sharing root systems. In such cases the original organism can be as much as 80,000 years old, as in the case of the Pando quaking aspen, found in Utah.

1,000,000

"*Who wants to be a millionaire? I don't.*
Have flashy flunkeys everywhere? I don't.
Who wants the bother of a country estate?
*A country estate is something I'd hate!***"**

"Who Wants to be a Millionaire" by Cole Porter,
from the film *High Society* (1956)

THE IDEA OF BEING a millionaire used to be beyond the wildest dreams of most of us. It was a status reserved for royalty or phenomenally successful businessmen. Today, just about everybody who owns their own business aspires to become a millionaire, if they're not one already. But they say the first million is the hardest to make. Of course, becoming a millionaire in the USA is easier than becoming one in the UK, since £1m is approximately twice the value of $1m at the time of writing. In Kuwait, the country with the world's most highly valued currency, it's even harder, the Kuwaiti dinar being worth about 3.5 US dollars. Nevertheless, millionaires abound in Kuwait thanks to the country's oil industry.

If we take a millionaire to be someone with the equivalent of $1m, there are close to 3 million millionaires in North America alone, and about 8 million worldwide.

■ **GAMESHOWS** like *Who Wants to Be a Millionaire?* have done their best to boost the world's millionaire population, although the contestants have rarely responded in kind. In the UK, where the programme was launched in September 1988, it took over two years for anyone to scoop the top prize. Judith Keppel, who is the third cousin of Camilla Parker Bowles (now the Queen Consort of the UK), was the first of five bona fide UK winners to date.

In 2001 Major Charles Ingram won the million-pound prize, but was later stripped of the money, fined £15,000 and handed an 18-month suspended jail sentence for fraud, along with his wife Diana, when it was found that they had cheated by using a system of coughs. Ingram appealed, but the verdict was upheld.

Who Wants to be a Millionaire? has been licensed to 69 countries.

ANOTHER WAY to get rich quick is to win a lottery. Nevertheless, in 2002 Valerie Wilson, a deli worker from New York, beat odds of 5,200,000–1 to scoop the Cool Million stake lottery scratch card jackpot of $1,000,000. Four years later, she won another million in the Jubilee scratch card game. This time the odds were 705,600–1, making the odds of winning the $1m jackpot in both games 3,669,120,000,000–1, or three trillion to one. Valerie said she had no intention of giving up her job.

■ The prefix **"mega"** used for measures of magnitude one million times does stem from a Greek word, *mega*, but its literal translation is simply "large" or "mighty." Similarly "micro," used for measures one millionth of the scale, is from the Greek *micros* meaning small.

Megawatt — a million watts

Megaton — measure of explosive force equal to one million tons of TNT

Megadeth — American thrash metal band

❏ THE WORD MILLION first came into use some time in the 13th century. It was derived from the Latin *mille* meaning a thousand, the word augmented in Italian to *millione*—"a great thousand." The Romans and Greeks had no word for million. If they ever wanted to express such a vast sum, they would have had to say "ten hundred thousand."

✱ *ONE MILLION YEARS BC* is a 1966 film about the exploits of prehistoric man. In reality, the hominids walking the earth at that time would not have looked like Raquel Welch, who memorably starred in the film.

10,000,000
etc.

AT LEAST 10 MILLION people worldwide have been to see *The Mousetrap* by Agatha Christie, the world's longest-running play. In 2022 it celebrated its 70th anniversary at St Martin's Theatre in London's West End. Christie wrote *The Mousetrap* in 1948 to celebrate the 80th birthday of Queen Mary. Originally called *Three Blind Mice*, it was only 30 minutes long and written for radio, but its adaptation for the stage somehow captured the public imagination.

121,174,811

is the first number in an extraordinary sequence of prime numbers, discovered in 1967 by L.J. Lander and T.R. Parkin. The sequence is as follows:

121,174,811
121,174,841
121,174,871
121,174,901
121,174,931
121,174,961

These six numbers are consecutive primes with a difference of 30 between each one.

3.9 billion

After the 2004 Olympic Games in Athens, IOC President J. Rogge announced some statistics. One of them was that 3.9 billion people had had access to the Games on television. Presumably he meant an American billion, as the total world population is about 6,500,000,000. That means 60 per cent of the world's population were able to watch the Olympics.

Global television audiences for any single event are dominated by sport. In 2006, the average audience for the FIFA World Cup final was more than 260 million.

To infinity...

BEYOND A HUNDRED MILLION, numbers start to become rather bewildering and, unless you're Elon Musk or Jeff Bezos, it's hard to find them having any bearing on everyday life. There is even disagreement as to what these monster numbers should be called. For example, a billion in the US is a thousand million; in Europe, it is a million million—a trillion in the US. There are some familiar examples of huge numbers, although you seldom stop to think about the quantities involved. You'll have heard of the gigabyte (1,000,000,000 bytes) in reference to the power of your PC; you may even have heard of the terabyte (1,000 gigabytes). You will also have heard of a nanosecond (a thousand-millionth of a second), probably while watching *Star Trek* or calling the IT department to fix your computer. "I'll be there in a nanosecond." Three weeks later...

4,320,000,000

❑ **4,320,000,000** years is one day in the life of Brahma, according to Hindu mythology. There are 1,000 *mahayuga* in one day of Brahma (*manvantara*) and 10 kali *yuga* (see **432**) in a *mahayuga*. One night of Brahma is also 4,320,000,000 years and a year of Brahma is 360 of his days and nights (3,110,400,000,000 years). The lifetime of Brahma (*maha-kalpa*) is 100 of these, and after two Brahman lifetimes the cycle of time starts all over again in an identical repeat of 622,080,000,000,000 years.

500,000,000,000

DURING THE BALKANS conflict in the early 1990s, Yugoslavia was gripped by massive inflation, forcing repeated revaluation of its currency. In 1993, a note was issued that bore the value 500,000,000,000 dinar.

■ THE BASIC DIFFERENCE between the US and the European numbering system for powers of 10 beyond a million is that a factor of a thousand (10^3) is inserted between each named number in the American system. These use the ending "-ard" instead of "-ion," so 1,000,000,000 is called a **milliard**. However, a thousand billion is usually called just that, to avoid confusion with the game of billiards. Not all of Europe uses this system. Greece, Italy, Russia and Turkey favor the American billion.

Number	US value	Europe value
Million	10^6	10^6
Billion	10^9	10^{12}
Trillion	10^{12}	10^{18}
Quadrillion	10^{15}	10^{24}
Quintillion	10^{18}	10^{30}
Sextillion	10^{21}	10^{36}
Septillion	10^{24}	10^{42}
Octillion	10^{27}	10^{48}
Nonillion	10^{30}	10^{54}
Decillion	10^{33}	10^{60}
Undecillion	10^{36}	10^{66}
Duodecillion	10^{39}	10^{72}
Tredecillion	10^{42}	10^{78}
Quattordecillion	10^{45}	10^{84}
Quindecillion	10^{48}	10^{90}
Sexdecillion	10^{51}	10^{96}
Septendecillion	10^{54}	10^{102}
Octodecillion	10^{57}	10^{108}
Novemdecillion	10^{60}	10^{114}
Vigintillion	10^{63}	10^{120}

These names continue up to **centillion**, which is 10^{303} in the US and 10^{600} in Europe. You'll notice that neither of these systems includes a name for **10^{100}**. It took a nine-year-old boy called Milton Sirotta to come up with that. He was the nephew of a mathematician called Edward Kasner, who asked him what he thought they should call a number that big. Milton said it should be called something silly like a "googol." And the name stuck.

Googol

The internet search engine Google took its name from Milton Sirotta's googol, but the different spelling came about as the result of a typo. When its founders, Larry Page and Sergey Brin, were checking to see whether the domain name was available, one of their students keyed it in as Google and found that it was. It struck a chord with Page and Brin so they went with it.

★

Of course, somebody had to go one better. What would you call a number that was 10 to the power of googol? Answer: a googolplex. That would be a 1 followed by 10^{100} zeros. It seems fitting that as zero was where we began this book, it should play such a major part in the huge numbers at the end.

602, 214,199,000,000,000,000,000

■ Otherwise known as **Avogadro's number** or the Loschmidt constant, 6.02×10^{23} is etched on the brain of every aspiring chemistry student (the exact number is constantly under review but this abbreviation is the one commonly used). In chemistryspeak, it is the number of elementary particles in a fixed mass of an element equal to the atomic weight of that element. It was first evaluated by the Austrian scientist, Johann Josef Loschmidt in 1865, but later named after Amedeo Avogadro, an Italian scientist. It enables you to calculate the number of particles in a given weight of any substance, based on its atomic weight.

$2^{32,582,657} - 1$

❏ IN 2006, this was discovered to be the largest known prime number. It's too big to write in digits without wasting paper. It consists of 9,808,358 digits, about 700,000 digits longer than the previous largest known prime.

It was discovered as part of a programme called the **Great Internet Mersenne Prime Search** launched in 1996 by a man called George Woltman. Only a relative handful of Mersenne Primes have been found, currently fewer than 50, although the GIMPS programme is throwing up bigger and bigger primes every year.

Space

299,792,458m/s
The speed of light

5,000
APPROXIMATE NUMBER OF STARS VISIBLE TO THE NAKED EYE

13,700,000,000 years:
Estimated age of the known Universe

Distance to Proxima Centauri, the nearest star to the Sun: **4.2 LIGHT YEARS**

5,879,000,000,000 miles
(9,460,730,472,580.8km)
A light year
(distance light travels in a vacuum in one year)

1,988,500,000,000,000,000,000,000,000,000,000g
MASS OF THE SUN

1,000,000,000,000,000,000,000,000,000,000W
POWER OUTPUT OF THE SUN

5,400,000,000,000,000,000,000,000,000,000g
MASS OF THE EARTH

70,000,000,000,000,000,000,000

The approximate number of stars in the universe, according to a
2004 study led by Dr Simon Driver of the
Australian National University

93,000,000 miles (149,598,000km)

= 1 Astronomical Unit (AU):
the distance from Earth to the Sun

The known Solar System

1 SUN	3 DWARF PLANETS
8 PLANETS	162 MOONS

	Distance from Sun (AU)	Rotation (days)	Orbit (days)	Moons
Mercury	0.4	58.65	87.66	0
Venus	0.7	243.02	226.45	0
Earth	1.0	1.00	365.24	1
Mars	1.5	1.03	686.65	2
Jupiter	5.2	0.41	4,331.75	63
Saturn	9.5	0.44	10,756.32	56
Uranus	19.6	0.72	30,687.46	27
Neptune	30.0	0.67	60,187.90	13

...and beyond

∞

> **❝**So, naturalists observe, a flea
> Hath smaller fleas that on him prey
> And these hath smaller fleas to bite 'em
> And so proceed ad infinitum.**❞**

So wrote Jonathan Swift, the Irish satirist who gave us *Gulliver's Travels*. In this poem he conveys the notion that life as we know it is replicated in smaller and smaller forms "ad infinitum"—to infinity. And so it might be in the other direction: that the world we live upon is a particle within a gigantic organism, which in turn is living on a particle... and so forth.

After reading about so many numbers, each with its own significance, each a measure of something finite, it is a challenge to the human brain to contemplate infinity, the great innumerable. But contemplate it we must, for while a number such as 10^{googol} can be easily put aside as a number that is effectively useless as anything other than a conversation topic, we find ourselves constantly grappling with concepts that require an understanding of infinity: **The universe—Time—A bottomless pit—A Moebius strip—Parallel lines—Death.**

One of the greatest breakthroughs of the Renaissance period in art was the mastery of perspective. This was achieved by applying a vanishing point—the point at which parallel lines appear to meet. Of course, parallel lines never meet. They go on side by side to infinity.

Consider this. An object approaching a fixed point is traveling at a constant speed such that, after one second, the distance is halved; after 1.5 seconds, the distance is halved again; after 1.75 seconds, it is halved again and so on. By this definition, the object will never actually reach the fixed point, because with each fraction of a second it only halves the distance remaining. Both the time and the distance can theoretically be split infinitely.

But don't spend too long contemplating infinity. It's enough to drive anyone mad, as well as a good point at which to bring to an end this infinitely broad subject of numbers.